MATHEMATICS FOR QUANTUM CHEMISTRY

JAY MARTIN ANDERSON
Franklin & Marshall College

DOVER PUBLICATIONS, INC.
Mineola, New York

With the fondest of memories, to my first class:

BARBARA, CANDY, ELLEN, LYNNE, MARCIA, ROBERTA, AND ROWENA

Copyright

Copyright © Copyright c 1966, 1974, by Jay Martin Anderson
Copyright © Renewed 1994, 2002 by Jay Martin Anderson
All rights reserved.

Bibliographical Note

This Dover edition, first published in 2005, is an unabridged reprinting of the corrected third printing (1974) of the work originally published in 1966 by W. A. Benjamin, Inc., New York and Amsterdam.

Library of Congress Cataloging-in-Publication Data

Anderson, Jay Martin.
  Mathematics for quantum chemistry / Jay Martin Anderson.
    p. cm.
  "This Dover edition, first published in 2005, is an unabridged republication of the work originally published in 1966 by W. A. Benjamin, Inc., New York and Amsterdam"—T.p. verso.
  Includes bibliographical references and index.
  ISBN 0-486-44230-6 (pbk.)
  1. Quantum Chemistry. 2. Chemistry—Mathematics. I. Title.

QD461.A5 2005
541'.28—dc22

2004061896

Manufactured in the United States of America
Dover Publications, Inc., 31 East 2nd Street, Mineola, N.Y. 11501

Preface

Recent trends in chemical education force students of chemistry into contact with the applications of quantum mechanics at an earlier and earlier point in their development. This direction of undergraduate chemistry curricula is important for several reasons: the wealth of physico-chemical information which is gleaned from molecular spectroscopy and other techniques which are based on quantum mechanics; the trend of descriptive chemistry to rely on concepts taken from quantum mechanics; the growing concern in theoretical chemistry to adequately explain molecular phenomena by means of quantum mechanics.

The rush to introduce the student to these bold and exciting studies often leaves him uncomfortable in the mathematical formalism in which these studies are couched, unsure of the connection between the conventional Newtonian world to which he is accustomed and the world of molecular phenomena, and unable to apply his newfound theoretical knowledge to problems of molecular structure and motion.

In the present work I shall attempt to present only two main topics from mathematics, and two from physics. These are the calculus of orthogonal functions and the algebra of vector spaces from mathematics, and the Lagrangian and Hamiltonian formulation of classical mechanics and its applications to molecular motion from physics. I have selected these four topics because of their relevance to modern quantum chemistry, especially in the application of quantum mechanics to molecular spectroscopy. This emphasis on molecular spectroscopy betrays my personal interest and excitement in this growing and popular field of endeavor; it also eliminates from the pages of this brief book a consideration of other topics which may be equally stimulating to my colleagues and to their students. Relativity, electricity, magnetism, and radiation physics were eliminated because they are generally better treated elsewhere and in greater depth than this work allows; similarly, group theory and differential equations, including approximate methods of solution, are left to other treatises.

This book attempts to lay down a central core of physical and mathematical background for quantum chemistry in general, but for molecular spectroscopy in particular. It assumes a knowledge of calculus through partial derivatives and multiple integration (usually about one and one-half years), a year of physics, and chemistry through a year of physical chemistry. This material has been used as the basis of a one-semester course at Bryn Mawr College entitled "Applied Mathematics for Chemists" for students with approximately the indicated background; this course immediately precedes the first course in quantum mechanics.

The author is indebted to Addison-Wesley Publishing Co. for permission to quote from their publications, and W. A. Benjamin, Inc. for their continued help and encouragement.

JAY MARTIN ANDERSON

Bryn Mawr, Pennsylvania
October 1965

Contents

Preface, iii

1 Introduction, 1

1–1	EIGENVALUE PROBLEMS IN QUANTUM MECHANICS	1
1–2	EIGENVALUE PROBLEMS IN CLASSICAL MECHANICS	4
1–3	SCOPE OF THIS BOOK	4
PROBLEM		5

2 Orthogonal Functions, 6

2–1	INTRODUCTORY CONCEPTS: ORTHOGONALITY AND NORMALIZATION	6
2–2	EXPANSION IN TERMS OF ORTHONORMAL FUNCTIONS	15
2–3	THE FOURIER SERIES	21
2–4	CONSTRUCTION OF ORTHONORMAL FUNCTIONS	27
2–5	THE LEGENDRE POLYNOMIALS AND OTHER SPECIAL FUNCTIONS	32
PROBLEMS		45

Contents

3 Linear Algebra, 48

3–1	INTRODUCTION	48
3–2	MATRICES, DETERMINANTS, AND LINEAR EQUATIONS	57
3–3	LINEAR TRANSFORMATIONS	76
3–4	LINEAR OPERATORS	84
PROBLEMS		101

4 Classical Mechanics, 105

4–1	INTRODUCTION AND THE CONSERVATION LAWS	105
4–2	GENERALIZED COORDINATES AND LAGRANGE'S EQUATIONS; HAMILTON'S EQUATIONS	110
4–3	VIBRATIONS OF A MECHANICAL SYSTEM	121
4–4	ROTATIONS OF A RIGID MECHANICAL SYSTEM	129
PROBLEMS		134

5 Conclusion, 136

5–1	THE BRIDGE BETWEEN CLASSICAL AND QUANTUM MECHANICS	136
5–2	THE SYNTHESIS OF MATRIX AND WAVE MECHANICS	140
PROBLEMS		142

Appendix Mathematical Background and Bibliography, 143

A-1	COMPLEX NUMBERS	143
A-2	CALCULUS: PARTIAL DERIVATIVES	144
A-3	BIBLIOGRAPHY	145

Answers to Problems, 147

Index, 151

1

Introduction

1–1 EIGENVALUE PROBLEMS IN QUANTUM MECHANICS

The mathematics and physics that are relevant to quantum chemistry are, almost without exception, oriented toward the solution of a particular kind of problem, the calculation of properties of a molecular system from the fundamental properties (charge, mass) of the particles composing the system. A good example of this is the calculation of the *energy* of the electrons in a molecule, using only the charge of the electron, Planck's constant, and so forth. The reader is probably already aware of the nature of the answer to this problem. There are a number of discrete values for the energy which the electrons in the molecule can assume up to a point, but higher values for the electronic energy occur in a continuous range. These energy values are shown qualitatively in Fig. 1–1. Quantum mechanics does provide the result that some physical quantities may take on only *some* values, not *all* values, as experiments indicated. The allowed values for a physical quantity are called *eigenvalues*, from the German for *characteristic values*. A particular physical quantity may assume an eigenvalue from a continuum, or perhaps from a finite or infinite discrete set of eigenvalues. The energy of an atom, for instance, may take on one of an

1

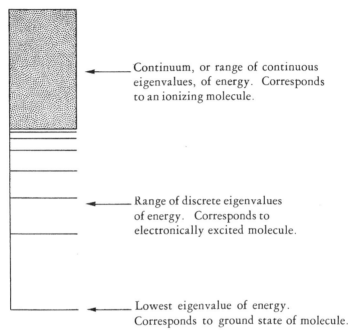

Figure 1-1 **Eigenvalues of the energy of a molecule.**

infinite number of discrete values, as well as values from a higher-lying range of continuous eigenvalues, called the *continuum*. More often than not, chemistry is concerned with the discrete eigenvalues of a quantity, rather than its continuum of eigenvalues.

The mathematical problem of finding the eigenvalues of a quantity is called an *eigenvalue problem;* it is usually cast in the form of an equation called an *eigenvalue equation*. An eigenvalue equation for a physical quantity Q has the deceptively simple appearance

$$Qf = qf \qquad (1\text{-}1)$$

In this equation, f is a function, called the *eigenfunction* for the quantity Q, with the eigenvalue q. The element Q is called an *operator*, and the statement Qf tells us to change the function f into a new function, according to a set of instructions implicit in the definition of the operator Q. The eigenvalue equation, Eq. 1-1, then informs

Introduction 3

us that, by applying these "instructions" of the operator Q to f, we get merely a multiple, q, of the function f. The function Qf differs from the function f by a multiplicative constant q. It may very well be the case that several eigenfunctions have the same eigenvalue; that is, $Qf_1 = qf_1$, $Qf_2 = qf_2$, and so forth. If this is the case, the eigenvalue q is said to be *degenerate*, and the number of eigenfunctions that have the same eigenvalue is called the *degree of degeneracy*.

Operators may simply be numbers or functions; for example, the operator $\mathfrak{X}$ may be defined by the instruction "multiply the operand function by x"; thus, $\mathfrak{X} x^2 = x^3$. On the other hand, operators may be more complex than just numbers or functions. For example, the student has already used the operator (although probably not by that name) Δ, which means, or is defined by the instructions, "evaluate the change in." For example, if we operate Δ on the thermodynamic function H, the enthalpy, we get a new function ΔH, the change in the enthalpy, $\Delta H = H_2 - H_1$. Another operator that is familiar is d/dx, meaning, "evaluate the derivative with respect to x."

It is the job of quantum mechanics to tell us how to form operators corresponding to the physical quantities which we wish to measure. Our task for the moment will be to learn how to solve the eigenvalue equations for such operators, and especially the vocabulary and concepts that are used to discuss the solutions. Quantum mechanics itself, however, grew up from two different points of view, which represent two analogous mathematical formulations of eigenvalue problems.

The first of these points of view is the *wave mechanics* of Schrödinger In wave mechanics, operators are differential expressions, such as the operator d/dx referred to above, and the eigenvalue equation then takes the form of a differential equation, and relies on the calculus for its solution. The second formulation is the *matrix mechanics* of Heisenberg, in which operators are represented by algebraic entities called *matrices;* instead of a function in the eigenvalue equation, the matrix operator operates on a *vector* ξ to transform ξ into a vector parallel to ξ, but q times as long:

$$Q\xi = q\xi \qquad (1\text{-}2)$$

Equation 1-2 is the matrix-mechanical formulation of the eigenvalue problem. Matrices and vectors are defined and discussed in detail in Chapter 3. As in Eq. 1-1, q is the eigenvalue of the quantity Q, ξ is the *eigenvector*, and Q is the operator represented as a matrix. The solution of this form of the eigenvalue problem relies on algebra. These apparently different mathematical and physical approaches to quantum mechanical problems are really deeply interrelated; the work of Dirac shows the underlying equivalence of the two points of view, as well as of the corresponding mathematical techniques.

1-2 EIGENVALUE PROBLEMS IN CLASSICAL MECHANICS

We have briefly discussed the role of eigenvalue equations in quantum mechanics. But a number of problems of classical mechanics may also be expressed in a simple and meaningful way as eigenvalue problems. Among these are the problems of the vibrations and rotations of a mechanical system, such as a molecule. These physical problems are of importance to the chemist concerned with molecular motion and spectroscopy. In vibrations, the normal modes and frequencies of oscillation appear as eigenvectors and eigenvalues; in rotations, the principal axes and moments of inertia emerge from an eigenvalue problem. It should be noted, however, that a correct description of these systems on the molecular level nearly always requires quantum mechanics, not classical mechanics.

1-3 SCOPE OF THIS BOOK

With our course thus determined by the kinds of problems we wish to be able to set up, solve, and understand, we shall proceed first to a study of a certain class of functions germane to eigenfunction problems, then to a number of aspects of vector algebra and matrix algebra, finally to a synthesis of the two points of view of eigenvalue problems. We shall conclude with a study of classical mechanics to see how the vibrations of a mechanical system, such as a molecule, may be formulated as an eigenvalue problem. We shall also attempt to formulate Newtonian mechanics in such a way that the connection to quantum mechanics is clear.

Introduction

Along the way, we shall learn some methods of solving eigenvalue problems, and take up applications of interest in chemistry. Our emphasis throughout will be primarily on concepts, secondarily on methods, and only lastly on the detailed proofs of the mathematical theorems. At the end of each chapter, a set of problems is given. Answers and hints for solution for many of the problems are found at the back of the book.

Problem

1. Find the eigenfunctions of the operator d/dx.

2

Orthogonal Functions

Two properties are, almost without exception, possessed by the eigenfunctions of operators corresponding to important physical quantities: *orthogonality* and *normality*. The purpose of this chapter is to develop these concepts in detail and to illustrate a number of their applications. Of primary usefulness is the idea of an *expansion in orthogonal functions*. As an example of this technique, we shall examine the *Fourier series* in some detail. We shall also learn how to construct orthogonal functions by the *Schmidt orthogonalization procedure*, and how orthogonal functions arise from the solution of particular differential equations. To illustrate the latter concepts, we shall investigate the properties of the *Legendre polynomials*, and briefly mention other of the important "special functions" which arise in quantum chemistry. A brief discussion of some of the elements of the calculus and of complex variables are given in the Appendix. The reader would be wise to check his familiarity with this material before advancing into the present chapter.

2-1 INTRODUCTORY CONCEPTS: ORTHOGONALITY AND NORMALIZATION

We may best begin our discussion of orthogonal functions by reviewing the concept of *function*. The concept of function has three essential ingredients. We agree first to define a function on a particu-

Orthogonal Functions

lar region of the number scale, say, from a to b. Second, we agree that there exists a variable (say, x) that can independently assume values in the region from a to b. Third, we agree by some prescribed rule that for any value of x there exists a *definite* value of y. Then we say that y is a function of x on the range $a \leq x \leq b$. This definition may be modified in a number of ways—so as to include more than one independent variable—but these three essential ingredients persist: an independent variable; a range on which the independent variable assumes its values; a dependent variable related to the independent variable by a prescribed rule.

The simplest way of notating the statement "y is a function of x" is to write $y = y(x)$. This notation is compact, yet may be misleading. The left side of the equation is simply the name of a variable—we do not know it is the dependent variable until we see the right side of the equation. The right side uses the letter y again, but here the symbol $y(\ )$ means something different than just the name of the variable. The meaning of $y(\ )$ is that y is a dependent variable whose value may be found by some prescribed rule from the quantity inside the parentheses. Left out of the notation $y = y(x)$ is the interval, or range, of the independent variable x for which the functional relationship is defined. This is not always of importance in elementary considerations of the idea of function, but it is of supreme importance to the notion of expansion of a function.

Hence, we introduce a definition.

> **Definition** *Expansion interval* (or, simply, *interval*). The expansion interval is the range of the independent variable assumed by the functions under consideration. This does not imply that the function may not be defined for other values of the independent variable; we just decline to consider those other values.

The expansion interval is usually notated $[a, b]$, meaning that the independent variable x is allowed values on the range $a \leq x \leq b$.

We proceed now to four definitions in rapid succession.

> **Definition** *Inner product*. The inner product of the two (in general, complex-valued) functions f and g of a continuous variable on their expansion interval $[a, b]$ is
>
> $$\langle f \mid g \rangle = \int_a^b f(x)^* g(x)\, dx \qquad (2\text{-}1)$$

The inner product of two functions is defined *on their expansion interval*. The inner product is notated by some authors (f, g), but this can easily be confused with the notation for two-dimensional coordinates or for an open interval. We shall use the notation $\langle f \mid g \rangle$. The order is quite important:

$$\langle g \mid f \rangle = \int g(x)^* f(x)\, dx = \left(\int f(x)^* g(x)\, dx \right)^*$$
$$= \langle f \mid g \rangle^* \qquad (2\text{-}2)$$

For real-valued functions, the order is not important. Equation 2-2 illustrates an important feature of the inner product that arises again and again: "turning around," or *transposing an inner product gives the complex conjugate of that inner product*. Constants may be removed at will from the inner product symbol: if b and c are (complex) numbers, $\langle bf \mid cg \rangle = b^* c \langle f \mid g \rangle$.

The inner product is a concept of no small significance. It has a geometrical analog, that of the *dot product* or *scalar product* of vectors that may already be familiar, which we shall discuss in Chapter 3.

In analogy to the geometrical property of perpendicularity of vectors, both functions and vectors afford the sweeping and general concept of *orthogonality*.

Definition Two functions, $f(x)$ and $g(x)$, are said to be *orthogonal* on the interval $[a, b]$ if their inner product on $[a, b]$ is zero:

$$\langle f \mid g \rangle = \int_a^b f^* g = 0 = \int_a^b g^* f = \langle g \mid f \rangle \qquad (2\text{-}3)$$

If the inner product is to be zero, it does not matter which function "comes first" in the inner product, so the orthogonality of f and g may be expressed by either $\langle f \mid g \rangle = 0$ or $\langle g \mid f \rangle = 0$. The perpendicularity of two vectors is related to this definition of orthogonality: two vectors are perpendicular if their dot product is zero.

Definition The *norm* of a function on the interval $[a, b]$ is the inner product of the function with itself, and may be symbolized by N:

$$N(f) = \langle f \mid f \rangle = \int_a^b f^* f \qquad (2\text{-}4)$$

Orthogonal Functions

The norm of a function is a real, positive quantity; it is analogous to the square of the length of a vector. That the norm is real and positive may be easily demonstrated by

$$f^*f = (\operatorname{Re} f - i \operatorname{Im} f)(\operatorname{Re} f + i \operatorname{Im} f) = (\operatorname{Re} f)^2 + (\operatorname{Im} f)^2 \quad (2\text{-}5)$$

which is positive definite. Then the integral of f^*f, which gives the norm of f, is also positive definite. The positiveness of the norm is of use to us at once.

Definition A function is said to be *normalized*[1] if its norm is one; that is, if $\langle f | f \rangle = 1$.

Since the norm of a function on a particular interval is always simply a positive real number, we can always form a multiple of a given function which is normalized. Suppose f has a norm N. Then the function $f/N^{1/2}$ will have a norm of one, since

$$\left\langle \frac{f}{N^{1/2}} \middle| \frac{f}{N^{1/2}} \right\rangle = \frac{1}{N} \langle f | f \rangle = \frac{N}{N} = 1 \quad (2\text{-}6)$$

The process of dividing a function by the square root of its norm is called *normalizing the function*, or, sometimes, normalizing the function to unity.

Let us use the five definitions we have introduced thus far in some examples. Suppose we consider functions defined on the interval $[-1, 1]$. As an example of the computation of an inner product, let us evaluate $\langle x | x^2 \rangle$.

$$\langle x | x^2 \rangle = \int_{-1}^{+1} x^* x^2 \, dx = \int_{-1}^{+1} x^3 \, dx = \left. \frac{x^4}{4} \right|_{-1}^{+1} = 0 \quad (2\text{-}7)$$

The computation of this simple inner product gives zero. We therefore may state that, on $[-1, 1]$, x and x^2 are orthogonal functions. Notice the importance of specifying the interval: on the interval $[0, 1]$, the inner product $\langle x | x^2 \rangle$ is

$$\langle x | x^2 \rangle = \int_0^1 x^3 = \left. \frac{x^4}{4} \right|_0^1 = \frac{1}{4} \quad (2\text{-}8)$$

[1] Normalization to unity is not the only possible normalization, but it is the most common, and will be used throughout this book.

and the functions are not orthogonal. The expansion interval must be specified before a statement about orthogonality can be made. The same is true for normality. On $[-1, 1]$, the function x has the norm

$$N(x) = \langle x \mid x \rangle = \int_{-1}^{+1} x^2 = \tfrac{2}{3} \qquad (2\text{-}9)$$

but on the interval $[0, 1]$, the norm

$$N(x) = \langle x \mid x \rangle = \int_{0}^{1} x^2 = \tfrac{1}{3} \qquad (2\text{-}10)$$

One very useful property of functions may be introduced at this point. Very often, the integrals which form inner products may be simplified by using symmetry properties of the functions. This symmetry may be expressed by two definitions.

> **Definition** An *even function* is a function for which $f(x) = f(-x)$; an *odd function* is a function for which $f(x) = -f(-x)$.

Evenness or oddness is easily pictured. Figure 2–1a shows a graph of the function $f(x) = x^2$, which is even, since $(x)^2 = (-x)^2$. Graphically speaking, the plot of $f(x)$ is symmetrical about the ordinate axis. Figure 2–1b shows the function $f(x) = x^3$, which is odd, since $(x)^3 = -(-x)^3$. The plot to the right is the negative of the plot to the left of the ordinate axis. The integrals of even or odd functions are especially simple *if the interval is symmetric*. The following theorem results.

> **Theorem** The integral of an even function on a symmetric interval is twice the integral on the half-interval; the integral of an odd function on a symmetric interval is zero.

This theorem is illustrated graphically in Fig. 2–2. It may be proven by dividing the full symmetric interval into two half-intervals:

$$\int_{-a}^{a} (\text{even}) = \int_{-a}^{0} (\text{even}) + \int_{0}^{a} (\text{even}) \qquad (2\text{-}11)$$

But, since an even function of x is the same as the even function of $-x$, we may replace the integral over the negative half-interval

Orthogonal Functions   11

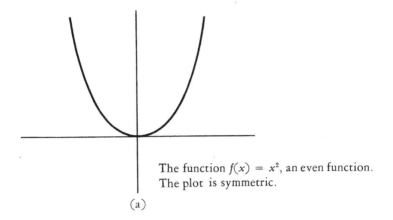

The function $f(x) = x^2$, an even function. The plot is symmetric.

(a)

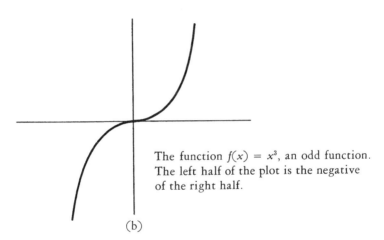

The function $f(x) = x^3$, an odd function. The left half of the plot is the negative of the right half.

(b)

Figure 2-1 Even and odd functions.

$[-a, 0]$ by the integral over the positive half-interval $[0, a]$ without changing the function:

$$\int_{-a}^{a} (\text{even}) = \int_{0}^{a} (\text{even}) + \int_{0}^{a} (\text{even}) = 2\int_{0}^{a} (\text{even}) \quad (2\text{-}12)$$

which proves the first part of the theorem. The second part is as simple:

$$\int_{-a}^{a} (\text{odd}) = \int_{-a}^{0} (\text{odd}) + \int_{0}^{a} (\text{odd})$$

$$= -\int_{0}^{a} (\text{odd}) + \int_{0}^{a} (\text{odd}) = 0 \qquad (2-13)$$

For odd functions, the integral over the negative half-interval may be replaced by an integral over the positive half-interval if the sign of the function is changed. Equation 2-13 gives the second part of the theorem.

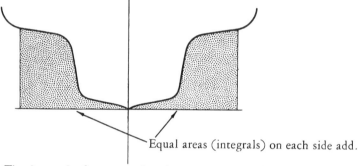

(a) The integral of an even function on a symmetric interval is twice the integral on the half-interval.

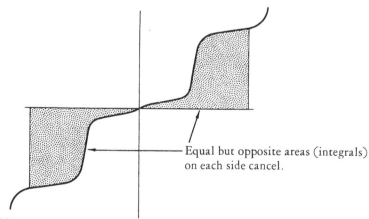

(b) The integral of an odd function on a symmetric interval is zero.

Figure 2-2 Integrals of even and odd functions.

Orthogonal Functions

Application of this theorem to the calculation of inner products on symmetric intervals gives another result.

Theorem On the symmetric interval $[-a, a]$, the inner product of an odd function with an odd function, or of an even function with an even function, is not zero, and may be calculated by twice the inner product over either half-interval $[-a, 0]$ or $[0, a]$; the inner product of an even function with an odd function is zero, no matter what the form of the function.

In the foregoing five definitions, we have considered only two arbitrary functions. However, the power and usefulness of the definitions established above really become apparent when we consider *sets of functions*. A set of functions is a collection of functions of the *same variable*, defined on the *same interval*, and *characterized by a rule* for figuring out all the functions in the set. For example, all the powers of x constitute a set of functions. We write this set, using braces, as $\{x^n\}$, indicating all these functions of x: $x^0 = 1$, $x^1 = x$, x^2, x^3, x^4, and so on. The general exponent n indicates the rule for figuring out each member of the set of functions. To be complete, we must specify the interval under consideration, and the values that n, often called the *index*, may take on, like this: "the set of functions $\{x^n\}$, on $[-1, 1]$, for $n = 0$ and all positive integers."

To round out our description of sets of functions which are useful in quantum chemistry, we introduce three new definitions.

Definition A *complete set of functions*, $\{F_i\}$, is a set of functions such that any other function f may be expressed as a linear combination of members of the set $\{F_i\}$ on a prescribed expansion interval to whatever precision may be desired.[2]

If the set $\{F_i\}$ is complete, then we may expand f in terms of the functions F_i like this:

$$f(x) = a_1 F_1(x) + a_2 F_2(x) + \cdots + a_n F_n(x) + \cdots$$
$$= \sum_{n=1}^{\infty} a_n F_n(x) \qquad (2\text{-}14)$$

---

[2] The uniform continuity of the functions of the set $\{F_i\}$ and of the function f is tacitly assumed here, and throughout; this definition could be made more rigorous if such concepts were used, but this is not of importance in quantum chemistry.

Generally speaking, proving that a set of functions is complete is quite difficult, and for our purposes we shall consider only such sets of functions which are known to be complete, and not consider completeness proofs per se.

> **Definition** An *orthogonal set of functions*, or a *set of orthogonal functions*, is a set of functions each of which is orthogonal on a prescribed interval to all other members of the set.

That is, the set $\{F_i\}$ is an orthogonal set if every member is orthogonal to every other member:

$$\langle F_j | F_k \rangle = 0 \qquad (2\text{-}15)$$

for all j and k such that $j \neq k$. Such a set of functions, which we shall consider in the third section of this chapter, is the set $\{\sin nx, \cos nx\}$, on $[-\pi, \pi]$, for n zero or positive integers. The proof of the orthogonality of these functions is one of the problems at the end of this chapter; notice that it involves three separate proofs:

$$\langle \sin nx | \sin mx \rangle = 0 \qquad n \neq m \qquad (2\text{-}16a)$$

$$\langle \cos nx | \cos mx \rangle = 0 \qquad n \neq m \qquad (2\text{-}16b)$$

$$\langle \sin nx | \cos nx \rangle = 0 \qquad \text{for all } n \qquad (2\text{-}16c)$$

Finally, we combine the definitions of orthogonality and normalization.

> **Definition** An *orthonormal set of functions* is an orthogonal set of functions, each of which is normalized.

That is, the set $\{F_i\}$ is orthonormal if

$$\langle F_j | F_k \rangle = 0 \qquad \text{for all } j \neq k \qquad (2\text{-}17a)$$

and

$$\langle F_j | F_j \rangle = 1 \qquad \text{for all } j \qquad (2\text{-}17b)$$

The pair of equations 2-17a and 2-17b occurs so often in discussing orthonormal functions that a special symbol has been introduced to combine Eqs. 2-17a and 2-17b. The *Kronecker delta symbol*, or *Kronecker delta*, or *delta symbol* δ_{ij} has the meaning $\delta_{ij} = 0$ for $i \neq j$, $\delta_{ij} = 1$ for $i = j$. If we use the Kronecker delta symbol, the condi-

Orthogonal Functions

tion for orthonormality, Eqs. 2–17a and 2–17b may be simply expressed as

$$\langle F_j \mid F_k \rangle = \delta_{jk} \quad \text{for all } j \text{ and } k \qquad (2-18)$$

In this chapter, orthonormal sets will be indicated by lower case Greek letters, such as $\{\phi_i\}$.

In this section we have defined a number of terms of importance, inner product, orthogonality, norm and normalization, completeness, and orthogonal and orthonormal sets of functions; we have also used the property of evenness or oddness of functions to simplify integrals over symmetric intervals.

2–2 EXPANSION IN TERMS OF ORTHONORMAL FUNCTIONS

In this section we shall learn how to expand a given function, on a prescribed interval, in terms of a set of orthonormal functions. Since the operations that occur in the calculation which follows occur often in discussions of orthonormal functions and later in discussions of orthonormal vectors, the calculation is set aside from the text.

$$f(x) = \sum_i a_i \phi_i(x) \qquad (2-19)$$

$$\phi_j(x)^* f(x) = \sum_i a_i \phi_j(x)^* \phi_i(x) \qquad (2-20)$$

$$\int \phi_j(x)^* f(x)\, dx = \sum_i a_i \int \phi_j(x)^* \phi_i(x)\, dx \qquad (2-21)$$

$$\langle \phi_j \mid f \rangle = \sum_i a_i \langle \phi_j \mid \phi_i \rangle \qquad (2-22)$$

$$\langle \phi_j \mid f \rangle = \sum_i a_i \delta_{ji} \qquad (2-23)$$

$$\langle \phi_j \mid f \rangle = a_j \qquad (2-24)$$

Equation 2–19 shows the expansion that we desire to use to express $f(x)$ on a particular expansion interval (not specified here) as a linear combination of the members of the set of orthonormal functions $\{\phi_i\}$.

Equation 2-20 gives the result of multiplying each side of the equation by $\phi_j(x)^*$, the complex conjugate of some member of the set $\{\phi_i\}$. Equation 2-21 gives the result of integrating both sides of Eq. 2-20 over the expansion interval. However, the result of Eqs. 2-20 and 2-21 is simply the formation of the inner product of ϕ_j (on the left) with Eq. 2-19 (on the right).

The definition of orthonormality is used in Eq. 2-23 to replace $\langle \phi_j | \phi_i \rangle$ with the Kronecker delta δ_{ji}.

Lastly, the sum over i on the right side of Eq. 2-23 is evaluated. If this sum were written out, it would look like this:

$$\sum_i a_i \delta_{ji} = a_1 \delta_{j1} + a_2 \delta_{j2} + \cdots + a_j \delta_{jj} + \cdots + a_n \delta_{jn} + \cdots \quad (2\text{-}25)$$

All the delta symbols but one are identically zero. The only one that is not zero is $\delta_{jj} = 1$. However, then the sum gives

$$\sum_i a_i \delta_{ji} = a_j \delta_{jj} = a_j \cdot 1 = a_j \quad (2\text{-}26)$$

The use of the delta symbols, which in turn stems from the property of orthonormality, is the key to a simple evaluation of the coefficients a_i for the expansion in orthonormal functions. The evaluation of the sum in Eq. 2-23 follows a simple rule: *A sum involving the products of a Kronecker delta with other quantities "picks out" that term for which the subscripts of the Kronecker delta are identical; or only that term survives for which the subscripts are identical.*

Finally, Eq. 2-24 is a formula for calculating the expansion coefficients in an expansion of a function of a given interval in members of an orthonormal set, $a_j = \langle \phi_j | f \rangle$. The expansion coefficients may be complex numbers. The use of a letter subscript should not obscure the issue: if we needed a_1, we would evaluate $\langle \phi_1 | f \rangle$; if a_2, $\langle \phi_2 | f \rangle$; if a_{309}, $\langle \phi_{309} | f \rangle$, and so on. We shall work an example in depth in the following section.

We turn next to a question of practical importance. If an expansion in orthonormal functions is curtailed after a finite number of terms, what error is incurred? The answer to this question reveals a new property of the expansion coefficients: these coefficients mini-

Orthogonal Functions

mize the error of a curtailed expansion. Denote the error after taking n terms by M_n. This error is evaluated[3] by

$$M_n = \int |f(x) - \sum_{j=1}^{n} b_j \phi_j|^2 \, dx \qquad (2\text{-}27)$$

that is, by the area under a plot of the square of the absolute value of the residual as a function of x. The integral is taken over the expansion interval. The meaning of M_n is illustrated graphically in Fig. 2-3. Since $c^*c = |c|^2$, we may write

$$\begin{aligned} M_n &= \int \left[f^* - \sum_{j=1}^{n} b_j^* \phi_j^* \right] \left[f - \sum_{i=1}^{n} b_i \phi_i \right] \\ &= \langle f | f \rangle - \sum_{j=1}^{n} b_j^* \langle \phi_j | f \rangle - \sum_{i=1}^{n} b_i \langle f | \phi_i \rangle \\ &\quad + \sum_{i,\, j=1}^{n} b_j^* b_i \langle \phi_j | \phi_i \rangle \end{aligned} \qquad (2\text{-}28)$$

Notice that a different summation index is used in each factor in Eq. 2-28. Summation indices are often referred to as *dummy indices*, since their *name alone* confers no special meaning. However, they must often be distinguished *by name* with care. For example,

$$\left(\sum_i c_i \phi_i \right) \left(\sum_i c_i \phi_i \right)$$

might casually be written

$$\sum_i c_i^2 \phi_i^2,$$

but this is not what the product connotes. If the sums are written out, $(c_1\phi_1 + c_2\phi_2 + \cdots)(c_1\phi_1 + c_2\phi_2 + \cdots)$ is equal to $c_1^2\phi_1^2 + c_1 c_2 \phi_1 \phi_2 + c_2 c_1 \phi_2 \phi_1 + c_2^2 \phi_2^2 + \cdots$, which is quite different from

---
[3] This is the "least-squares" criterion for error. Others are also applicable. It should also be noted that the curtailed expansion will not be normalized if the total function is.

$$\sum_i c_i^2 \phi_i^2 = c_1^2 \phi_1^2 + c_2^2 \phi_2^2 + \cdots.$$

To ensure that the proper answer—the one with the cross terms—is

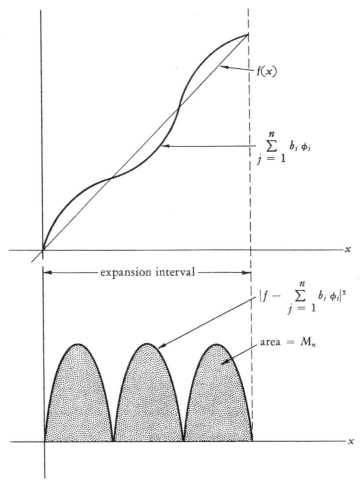

Figure 2-3 The error in a curtailed expansion in orthonormal functions is $M_n = \int |f - \sum_{j=1}^{n} b_j \phi_j|^2$.

Orthogonal Functions

written, we use the notation

$$\left(\sum_i c_i \phi_i\right)\left(\sum_j c_j \phi_j\right) = \sum_{i,j} c_i c_j \phi_i \phi_j.$$

In the last term of Eq. 2-28 we may substitute the Kronecker delta, because the set $\{\phi_i\}$ is orthonormal:

$$M_n = \langle f | f \rangle - \sum_{j=1}^n b_j^* \langle \phi_j | f \rangle - \sum_{i=1}^n b_i \langle f | \phi_i \rangle + \sum_{i,\,j=1}^n b_j^* b_i \delta_{ji}$$

$$= \langle f | f \rangle - \sum_{j=1}^n [b_j \langle f | \phi_j \rangle + b_j^* \langle \phi_j | f \rangle - b_j^* b_j] \quad (2\text{-}29)$$

The sum over i involving only the delta symbol leaves just the sum over j, with the terms surviving for which $i = j$, and, since the name of the summation index is unimportant, each of the three terms has been expressed in the index j, and combined.

We would find the "best" set of coefficients for the expansion by minimizing the error M_n with respect to these coefficients, b_j. This involves two steps since the b_j have both a real and an imaginary part. This double-minimum problem may be expressed by requiring $\partial M_n / \partial b_j = \partial M_n / \partial(b_j^*) = 0$ for all j. Taking derivatives of Eq. 2-29, we obtain

$$\frac{\partial M_n}{\partial b_j} = 0 = -\langle f | \phi_j \rangle + b_j^* \quad (2\text{-}30a)$$

$$\frac{\partial M_n}{\partial(b_j^*)} = 0 = -\langle \phi_j | f \rangle + b_j \quad (2\text{-}30b)$$

These equations are complex conjugates of one another, and the "best" coefficients for the expansion, in the sense of minimizing the error, are given by

$$b_j = \langle \phi_j | f \rangle \quad (2\text{-}31)$$

The coefficients given by Eq. 2-31 are, in fact, the coefficients that *must* be used in an expansion in orthonormal functions. Hence, $a_j = b_j = \langle \phi_j | f \rangle$ are not only the "right" coefficients (Eq. 2-24), but also the "best" (Eq. 2-31).

After substituting $a_j = b_j = \langle \phi_j | f \rangle$ into Eq. 2-29, we calculate the error to be

$$M_n = \langle f | f \rangle - \sum_{j=1}^{n} a_j a_j^* + a_j^* a_j - a_j^* a_j$$

$$= \langle f | f \rangle - \sum_{j=1}^{n} |a_j|^2 \qquad (2\text{-}32)$$

We conclude this section with an important relation concerning inner products, called the *expansion theorem*.

Theorem The inner product $\langle f | g \rangle$ may be expanded in terms of an orthonormal set $\{\phi_i\}$ as

$$\langle f | g \rangle = \sum_i \langle f | \phi_i \rangle \langle \phi_i | g \rangle \qquad (2\text{-}33)$$

To prove this theorem, expand f as $f = \sum_i a_i \phi_i$, and g as $g = \sum_j b_j \phi_j$. Then,

$$\langle f | g \rangle = \int f^* g = \sum_i \sum_j \int a_i^* \phi_i^* b_j \phi_j$$

$$= \sum_{ij} a_i^* b_j \langle \phi_i | \phi_j \rangle = \sum_{ij} a_i^* b_j \delta_{ij}.$$

However, $a_i^* = \langle \phi_i | f \rangle^* = \langle f | \phi_i \rangle$, and $b_i = \langle \phi_i | g \rangle$. Hence,

$$\langle f | g \rangle = \sum_i \langle f | \phi_i \rangle \langle \phi_i | g \rangle.$$

The structure $| \phi_i \rangle \langle \phi_i |$ which occurs in the expansion theorem will come to have more meaning later on in the book. For the time being, we may notice that in the expansion theorem we have but inserted $| \phi_i \rangle \langle \phi_i |$ between $\langle f |$ and $| g \rangle$ and summed over i. Such is indeed the case; this operation may be called "inserting a complete set of states." The content of the expansion theorem may be written $\sum_i | \phi_i \rangle \langle \phi_i | = 1$. The structure $| \phi_i \rangle \langle \phi_i |$ is not an inner product as

Orthogonal Functions

defined in Eq. 2–1; it is an element that has not heretofore entered our discussions. Actually, it is an operator. The sense in which it is an operator will become apparent later.

In this section, we have derived the formula for the coefficients in an expansion in terms of orthonormal functions; we have demonstrated that these coefficients minimize the error in a curtailed expansion, and we have derived a formula for the error that is incurred by curtailing an expansion. Finally, we have derived the expansion theorem for the inner product.

2–3 THE FOURIER SERIES

The Fourier series is an expansion of a function in the orthonormal functions which are proportional to $\{\sin mx, \cos mx\}$. We mentioned in the first section that these functions are orthogonal, but we derive now their norm on the interval $[-\pi, \pi]$.

$$N(\sin mx) = \langle \sin mx \mid \sin mx \rangle = \int_{-\pi}^{\pi} \sin^2 mx \, dx$$

$$= \frac{1}{m} \int_{-m\pi}^{m\pi} \sin^2 y \, dy = \pi \quad (2\text{–}34)$$

Similarly,

$$N(\cos mx) = \langle \cos mx \mid \cos mx \rangle = \frac{1}{m} \int_{-m\pi}^{m\pi} \cos^2 y \, dy = \pi \quad (2\text{–}35)$$

Equations 2–34 and 2–35 are valid for all values of m, except for $m = 0$, where each result yields the ambiguous $N = 0/0$. To clarify the $m = 0$ case, we must separately evaluate

$$N(\sin 0x) = \langle \sin 0x \mid \sin 0x \rangle = \int_{-\pi}^{\pi} 0 \, dx = 0 \quad (2\text{–}36)$$

and

$$N(\cos 0x) = \langle \cos 0x \mid \cos 0x \rangle = \int_{-\pi}^{\pi} 1 \, dx = 2\pi \quad (2\text{–}37)$$

All these results may be expressed by

$$N(\sin mx) = \pi \quad \text{for } m \neq 0$$
$$= 0 \quad \text{for } m = 0$$
$$N(\cos mx) = \pi \quad \text{for } m \neq 0$$
$$= 2\pi \quad \text{for } m = 0 \quad (2\text{-}38a)$$

or, using the Kronecker delta, by

$$N(\sin mx) = \pi - \pi\delta_{m0}$$
$$N(\cos mx) = \pi + \pi\delta_{m0} \quad (2\text{-}38b)$$

Together with the orthogonality relations, we may then write all possible inner products of $\{\sin mx, \cos mx\}$ by

$$\langle \sin mx \mid \cos mx \rangle = 0$$
$$\langle \sin mx \mid \sin nx \rangle = (1 - \delta_{m0})\pi\delta_{mn}$$
$$\langle \cos mx \mid \cos nx \rangle = (1 + \delta_{m0})\pi\delta_{mn} \quad (2\text{-}39)$$

for all $m, n =$ any positive integer or zero. Since we shall be using our previous formula, Eq. 2-24, for calculating the expansion coefficients for an expansion in orthonormal functions, we must use, instead of the set of functions $\{\sin mx, \cos mx\}$, the set of functions $\{(2\pi)^{-1/2}, \pi^{-1/2} \sin mx, \pi^{-1/2} \cos mx\}$, for $m = 1, 2, \ldots$, for expansion on the interval $[-\pi, \pi]$.

For these orthonormal functions, the expansion coefficients will be

$$a_0 = \langle (2\pi)^{-1/2} \mid f \rangle$$
$$a_m = \langle \pi^{-1/2} \cos mx \mid f \rangle$$
$$b_m = \langle \pi^{-1/2} \sin mx \mid f \rangle \quad (2\text{-}40)$$

where the expansion is

$$f(x) = a_0(2\pi)^{-1/2} + \sum_{m=1}^{\infty} a_m(\pi^{-1/2} \cos mx) + \sum_{m=1}^{\infty} b_m(\pi^{-1/2} \sin mx)$$
$$(2\text{-}41)$$

This is not the usual form of the Fourier series, but this is an example of an expansion in orthonormal functions. Usually, the constants are removed from the terms by explicitly writing out the expansion coefficients:

Orthogonal Functions

$$f(x) = \frac{1}{2\pi}\langle 1 | f \rangle + \frac{1}{\pi}\sum_{m=1}^{\infty}[\langle \cos mx | f \rangle \cos mx + \langle \sin mx | f \rangle \sin mx]$$

(2-42)

This gives the final result

$$f(x) = \frac{c_0}{2} + \sum_{m=1}^{\infty} c_m \cos mx + \sum_{m=1}^{\infty} d_m \sin mx \qquad (2\text{-}43)$$

$$c_m = \frac{1}{\pi}\langle \cos mx | f \rangle \qquad (2\text{-}44a)$$

$$d_m = \frac{1}{\pi}\langle \sin mx | f \rangle \qquad (2\text{-}44b)$$

on the interval $[-\pi, \pi]$. This is the form in which the Fourier series is usually written. Notice that the lead term is divided by two. This is because the norm of $\cos 0x$ is 2π, whereas for all other values of m, the norm of $\cos mx$ is π.

We can make two simple extensions of the Fourier series at once by using the property of evenness and oddness of functions. The sine function is odd; $\sin x = -\sin(-x)$. The cosine function is even; $\cos x = \cos(-x)$. This being the case, we may formulate these two rules.

(1) *The Fourier expansion on $[-\pi, \pi]$ of an odd function is made up only of sine terms:* $f(x) = \sum_{m=1}^{\infty} d_m \sin mx.$ (2) *The Fourier expansion on $[-\pi, \pi]$ of an even function is made up only of cosine terms:* $f(x) = c_0/2 + \sum_{m=1}^{\infty} c_m \cos mx.$ For, if f is odd, all inner products $\langle \cos mx | f \rangle \equiv 0$; and, if f is even, all inner products $\langle \sin mx | f \rangle \equiv 0$.

EXAMPLE

The expansion of $f(x) = x$ on $[-\pi, \pi]$. Since $f(x) = x$ is odd, only sine terms occur, and we may write

$$x = \sum_{m=1}^{\infty} d_m \sin mx \qquad m = 1, 2, \ldots \text{ on } [-\pi, \pi]$$

where

$$d_m = \frac{1}{\pi} \langle \sin mx \mid x \rangle$$

$$= \frac{1}{\pi} \int_{-\pi}^{\pi} x \sin mx \, dx = \frac{1}{m^2\pi} \int_{-m\pi}^{m\pi} y \sin y \, dy$$

$$= \frac{1}{m^2\pi} (-y \cos y + \sin y) \Big|_{-m\pi}^{m\pi}$$

$$= \frac{-m\pi \cos m\pi - m\pi \cos m\pi}{m^2\pi} = \frac{2(-1)^{m+1}}{m}$$

and the Fourier series is

$$x = \sum_{m=1}^{\infty} \frac{2(-1)^{m+1}}{m} \sin mx$$

$$= 2 \sin x - \sin 2x + \tfrac{2}{3} \sin 3x - \cdots$$

The comparison of the Fourier series curtailed after three terms with the function itself is shown in Fig. 2-4. The mean-square error after taking three terms can be found using Eq. 2-32 as follows:

$$M_3 = \langle x \mid x \rangle - \sum_{j=1}^{3} |a_j|^2 \qquad (2\text{-}45)$$

According to Eqs. 2-40 and 2-44a, $a_j = \pi^{1/2} c_j$, so that

$$M_3 = \int_{-\pi}^{\pi} x^2 \, dx - 4\pi - 1\pi - \frac{4}{9}\pi = \frac{2\pi^3}{3} - \frac{49\pi}{9} \simeq 6.8;$$

the square root of this, 2.6, may be compared with the area under the function $\pi^2 = 9.9$. Therefore, after three terms, a relative error of about 26% persists.

The use of Fourier series to describe functions has an important electronic analog. It is usually the case that one can generate sine and cosine functions electronically with ease. The generation of other functions can then be done by an electronic Fourier synthesis. Such a synthesis could be used to generate the so-called "sawtooth" wave shown in Fig. 2-5a. The sawtooth pattern is an endless series of plots of $f(x) = x$ vs. x on $[-\pi, \pi]$, which has the Fourier com-

Orthogonal Functions

ponents derived in the example. Figure 2-5b shows a three-term approximation to the sawtooth wave. Such waveforms are used for sweeps, as in oscilloscopes and television.

The functions $\{\sin mx\}$ and $\{\cos mx\}$ are individually complete on either half-interval $[-\pi, 0]$ or $[0, \pi]$. These functions are also orthogonal on either half-interval, and all have the norm $\pi/2$, except $\cos 0x$, whose norm is π. Therefore, we can construct two distinct "half Fourier series":

$$f(x) = \frac{c_0}{2} + \sum_{m=1}^{\infty} c_m \cos mx \qquad (2\text{-}46a)$$

$$c_m = \frac{2}{\pi} \langle \cos mx \mid f \rangle \qquad (2\text{-}46b)$$

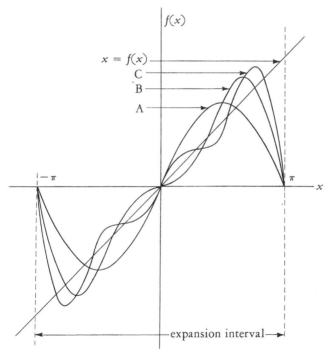

Figure 2-4 The function $f(x) = x$ on the interval $(-\pi, \pi)$, and the Fourier series for $f(x)$ after (A) one term, (B) two terms, and (C) three terms.

$$f(x) = \sum_{m=1}^{\infty} d_m \sin mx \quad (2\text{-}47a)$$

$$d_m = \frac{2}{\pi} \langle \sin mx \mid f \rangle \quad (2\text{-}47b)$$

each on either interval $[-\pi, 0]$ or $[0, \pi]$. Likewise, either the full or half Fourier series may be extended to any symmetric interval $[-a, a]$ or half of a symmetric interval $[-a, 0]$ or $[0, a]$ by a scale expansion, such as

$$f(x) = \frac{c_0}{2} + \sum_{m=1}^{\infty} \left(c_m \cos \frac{m\pi x}{a} + d_m \sin \frac{m\pi x}{a} \right) \quad (2\text{-}48)$$

with $c_m = (1/a)\langle \cos(m\pi x/a) \mid f \rangle$, $d_m = (1/a)\langle \sin(m\pi x/a) \mid f \rangle$. A very important revision of the Fourier series for quantum mechanics is the formulation

$$f(x) = \sum_{m=-\infty}^{\infty} a_m e^{imx} \quad \text{on } [-\pi, \pi] \quad (2\text{-}49)$$

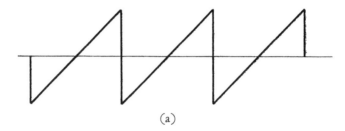

(a)

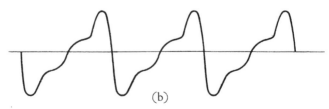

(b)

Figure 2-5 (a) The "sawtooth" wave, and (b) a three-term Fourier synthesis of the sawtooth wave.

Orthogonal Functions

For, if $\{\cos mx, \sin mx\}$ is complete for all positive values of m and zero, then $\{e^{imx} = \cos mx + i \sin mx\}$ is complete for all integral values—positive, negative, and zero—of m. To put it another way, including the complex conjugate of the functions $\{e^{imx}\}$ completes the set. These modifications of the basic Fourier series are subjects of a number of the problems at the end of the chapter.

In this section, we have exhibited a very useful expansion in orthonormal functions, the Fourier series (expansion in sines and cosines), and have commented on its usefulness and extension.

2-4 CONSTRUCTION OF ORTHONORMAL FUNCTIONS

Thus far, we have discussed in general the properties of orthonormal functions, their use in series expansions, and the particular example of the Fourier series. We have pointed out that a sufficiently precise expansion can be achieved only if the set of functions, orthonormal or not, is complete. In this section, we shall show how an orthonormal complete set of functions may be formed from a set of complete functions. Before we take up the actual formation, we must introduce one aspect of completeness which plays an important rôle in this discussion and also in the work that follows. We begin with a definition.

Definition A set of functions is said to be *linearly independent* if none of the functions can be expressed as a linear combination of the rest. Similarly, a set of functions is said to be *linearly dependent* if at least one of the functions may be related to one or more of the other functions by a linear equation.

This definition may be expressed mathematically by saying that a set of functions $\{F_i\}$ is linearly independent if the identity

$$\sum_i c_i F_i = c_1 F_1 + c_2 F_2 + \cdots = 0 \qquad (2\text{-}50)$$

can be solved only with all the $c_i = 0$; the set $\{F_i\}$ is linearly dependent if this equation can be solved with at least one of the $c_i \neq 0$.

Take, for example, these two sets of functions:

$$F_1 = 1 \qquad G_1 = 1$$
$$F_2 = x \qquad G_2 = x$$
$$F_3 = 7x + 4 \qquad G_3 = x^2$$
$$F_4 = x^3 \qquad G_4 = x^3$$

The set $\{G_i\}$ is linearly independent. It is not possible for any linear combination of those functions to add to zero, as in Eq. 2-50. To put it another way, there is no linear relation connecting those four functions: the function $G_3 = x^2$ is not a linear combination of 1, x, and x^3. The set $\{F_i\}$ is linearly dependent. Function F_3 is a linear combination of F_1, F_2, and F_4; $F_3 = 7F_2 + 4F_1$. That is, Eq. 2-50 can be solved with some of the $c_i \neq 0$; $-4F_1 - 7F_2 + 1F_3 + 0F_4 = 0$.

A set of complete functions will always contain a linearly independent subset. This is an important statement that is offered without proof; it is the first step in generating a complete set of orthonormal functions.

As an example, consider the powers of x on the interval $[-1, 1]$. Taylor's theorem essentially guarantees that the set $\{x^n\}$, $n = 0$, 1, 2, $\cdots$ is complete. We have just seen that these functions are also linearly independent. Consider the sets of functions below. The set $\{F_i'\}$ is complete; the linearly independent set $\{F_i\}$ was formed by removing the linearly dependent member of $\{F_i'\}$. The set $\{G_i\}$ is also linearly independent and complete; it differs from $\{F_i\}$ only in the order of arrangement.

$\{F_i'\}$	$\{F_i\}$	$\{G_i\}$
1	1	1
x	x	x^2
$2x$	$\cdots$	x^4
x^2	x^2	$\cdots$
x^3	x^3	x
x^4	x^4	x^3
$\cdots$	$\cdots$	$\cdots$

Construction of a complete set of orthonormal functions can always be accomplished from a complete, linearly independent set of functions. The procedure for this construction is known as the *Schmidt orthogonalization procedure*. Rather than give a complete account of the procedure, we shall illustrate the idea, state the general

Orthogonal Functions

conclusion, and then work an illustrative example. Suppose we have a complete, linearly independent set of functions $\{f_i\}$, defined on a prescribed interval. We wish to form an orthonormal set $\{\phi_i\}$ from the $\{f_i\}$.

Step 1. Let ϕ_0 be f_0 normalized. That is, let the first ϕ function be simply proportional to the first f function. The norm of f_0 is $N_0 = \langle f_0 \mid f_0 \rangle$, so that $\phi_0 = f_0/N_0^{1/2}$.

Step 2. Let ϕ_1 be a linear combination of ϕ_0 and f_1 which is orthogonal to ϕ_0, and normalized. This can be done in general since the set $\{f_i\}$ is linearly independent. That is,

$$\phi_1 = N_1^{-1/2}(c\phi_0 + f_1) \qquad (2\text{-}51)$$

where

$$\langle \phi_0 \mid \phi_1 \rangle = 0 \qquad (2\text{-}52)$$

The orthogonality condition gives

$$c\langle \phi_0 \mid \phi_0 \rangle + \langle \phi_0 \mid f_1 \rangle = 0 \qquad (2\text{-}53a)$$
$$c = -\langle \phi_0 \mid f_1 \rangle \qquad (2\text{-}53b)$$

Then,

$$\phi_1 = N_1^{-1/2}(f_1 - \langle \phi_0 \mid f_1 \rangle \phi_0) \qquad (2\text{-}54)$$

and

$$N_1 = \langle f_1 \mid f_1 \rangle - \langle \phi_0 \mid f_1 \rangle \langle f_1 \mid \phi_0 \rangle - \langle \phi_0 \mid f_1 \rangle \langle \phi_0 \mid f_1 \rangle + \langle \phi_0 \mid f_1 \rangle^2$$
$$= \langle f_1 \mid f_1 \rangle - |\langle \phi_0 \mid f_1 \rangle|^2 \qquad (2\text{-}55)$$

This process is then continued until the complete, orthonormal set is generated. The general term is

$$\phi_k = N_k^{-1/2}\left(f_k - \sum_{j=0}^{k-1} \langle \phi_j \mid f_k \rangle \phi_j\right) \qquad (2\text{-}56)$$

with

$$N_k = \langle f_k \mid f_k \rangle - \sum_{j=0}^{k-1} |\langle \phi_j \mid f_k \rangle|^2 \qquad (2\text{-}57)$$

It should be clear from this discussion that functions will not be normalizable if their norm is not finite. Not all functions on all intervals are normalizable: the set $\{x^n\}$ on $(-\infty, \infty)$ is not, for example. The quantum mechanics of molecular structure is concerned almost exclusively with normalizable, or *square-integrable* functions.

As an example of the Schmidt process, we shall consider the functions $\{\exp[-x^2/2]x^n\}$ on the full line, $(-\infty, \infty)$. We begin by showing that these functions are normalizable.

$$\langle f_n | f_n \rangle = \int_{-\infty}^{\infty} \exp\left[\frac{-x^2}{2}\right] x^n \exp\left[\frac{-x^2}{2}\right] x^n \, dx$$

$$= \int_{-\infty}^{\infty} \exp[-x^2] x^{2n} \, dx \qquad (2\text{-}58)$$

Since the integrand is an even function, we may simplify the integral to

$$\langle f_n | f_n \rangle = 2 \int_0^{\infty} \exp[-x^2] x^{2n} \, dx = \int_0^{\infty} e^{-y} y^{(n-1/2)} \, dy \qquad (2\text{-}59)$$

which may successively be integrated by parts to give

$$\langle f_n | f_n \rangle = (n - \tfrac{1}{2})(n - \tfrac{3}{2}) \cdots (\tfrac{1}{2}) \int_0^{\infty} e^{-y} y^{-1/2} \, dy$$

$$= 2(n - \tfrac{1}{2})(n - \tfrac{3}{2}) \cdots (\tfrac{1}{2}) \int_0^{\infty} e^{-x^2} \, dx$$

$$= (2n - 1)(2n - 3) \cdots (3)(1)\pi^{1/2}/2^n \qquad (2\text{-}60)$$

where the known integral $\int_0^{\infty} e^{-x^2} \, dx = \dfrac{\pi^{1/2}}{2}$ has been substituted.

The norm N_n is finite for any n, so all functions in the set $\{\exp[-x^2/2]x^n\}$ are normalizable. It is the presence of the decaying exponential $\exp[-x^2/2]$ that renders the integral finite and the functions square integrable.

To construct an orthonormal set, we begin by normalizing $f_0 = \exp[-x^2/2]x^0 = \exp[-x^2/2]$. The norm is $\pi^{1/2}$, so we have $\phi_0 = \pi^{-1/4} \exp[-x^2/2]$.

The next function is found from Eqs. 2-54 and 2-55, which are

Orthogonal Functions

specific cases of Eqs. 2-56 and 2-57, respectively. In order to use these equations, we must first evaluate $\langle \phi_0 | f_1 \rangle$ and $\langle f_1 | f_1 \rangle$:

$$\langle \phi_0 | f_1 \rangle = \int_{-\infty}^{\infty} \exp\left[\frac{-x^2}{2}\right] \pi^{-1/4} \exp\left[\frac{-x^2}{2}\right] x \, dx = 0 \quad (2\text{-}61)$$

because the integrand is odd, and

$$\langle f_1 | f_1 \rangle = \pi^{1/2}/2 \quad (2\text{-}62)$$

from Eq. 2-60. Then

$$N_1 = \langle f_1 | f_1 \rangle - |\langle \phi_0 | f_1 \rangle|^2 = \pi^{1/2}/2 \quad (2\text{-}63)$$

and

$$\phi_1 = N_1^{-1/2}(f_1 - \langle \phi_0 | f_1 \rangle \phi_0) = 2^{1/2} \pi^{-1/4} \exp\left[\frac{-x^2}{2}\right] x \quad (2\text{-}64)$$

The reader should notice that all integrals of the form $\langle \phi_i | f_j \rangle$ where i is even and j odd, or vice versa, are zero for these sets of functions because the integrand is odd. This greatly simplifies the calculation.

The next function will be calculated according to Eqs. 2-56 and 2-57, as follows:

$$\begin{aligned}\phi_2 &= N_2^{-1/2}(f_2 - \langle \phi_0 | f_2 \rangle \phi_0 - \langle \phi_1 | f_2 \rangle \phi_1) \\ &= N_2^{-1/2}(f_2 - \langle \phi_0 | f_2 \rangle \phi_0)\end{aligned} \quad (2\text{-}65)$$

and

$$\begin{aligned}N_2 &= \langle f_2 | f_2 \rangle - |\langle \phi_0 | f_2 \rangle|^2 - |\langle \phi_1 | f_2 \rangle|^2 \\ &= \langle f_2 | f_2 \rangle - |\langle \phi_0 | f_2 \rangle|^2\end{aligned} \quad (2\text{-}66)$$

We need to evaluate $\langle f_2 | f_2 \rangle$ and $\langle \phi_0 | f_2 \rangle$. From Eq. 2-60,

$$\langle f_2 | f_2 \rangle = \tfrac{3}{4}\pi^{1/2} \quad (2\text{-}67)$$

and

$$\begin{aligned}\langle \phi_0 | f_2 \rangle &= \int_{-\infty}^{\infty} \pi^{-1/4} \exp\left[\frac{-x^2}{2}\right] \exp\left[\frac{-x^2}{2}\right] x^2 \, dx \\ &= \pi^{-1/4} \int_{-\infty}^{\infty} x^2 \exp[-x^2] \, dx = \pi^{-1/4} \frac{\pi^{1/2}}{2} \\ &= \frac{\pi^{1/4}}{2}\end{aligned} \quad (2\text{-}68)$$

Finally,

$$N_2 = \frac{3}{4}\pi^{1/2} - \frac{1}{4}\pi^{1/2} = \frac{\pi^{1/2}}{2} \tag{2-69}$$

$$\phi_2 = (\sqrt{2}\pi^{-1/4})\left\{\exp\left[\frac{-x^2}{2}\right]x^2 - \frac{\pi^{1/4}}{2}\pi^{-1/4}\exp\left[\frac{-x^2}{2}\right]\right\}$$

$$= \pi^{-1/4}\exp\left[\frac{-x^2}{2}\right]\frac{(2x^2 - 1)}{\sqrt{2}} \tag{2-70}$$

The set of orthonormal functions that would be generated in this way are all of the form $\pi^{-1/4}(1/\sqrt{2^n n!})\,H_n(x)\exp[-x^2/2]$, where $H_n(x)$ represents one of the polynomials

$$\begin{aligned}H_0(x) &= 1\\ H_1(x) &= 2x\\ H_2(x) &= 4x^2 - 2\end{aligned} \tag{2-71}$$

...

The even-numbered polynomials contain even powers of x only, and the odd-numbered polynomials contain odd powers of x only. These polynomials are proportional to the famous *Hermite polynomials*. The functions that we have constructed to be orthonormal on $(-\infty, \infty)$ from the set $\{\exp[-x^2/2]x^n\}$ are the solutions for the wave function of a quantum-mechanical harmonic oscillator. We have, of course, carried out our construction with no attention to any particular physical problem. That we accidentally—by this route—come across the solution to a physical problem emphasizes the significance and simplicity of the harmonic oscillator in quantum mechanics.

Thus, in this section, we have further developed the properties of functions with the concepts of linear independence and normalizability, and we have shown how a complete orthonormal set of functions can be constructed from a complete, linearly independent set. Lastly, we have begun crudely to relate the properties of orthonormal functions to the wave functions describing physical situations.

2-5 THE LEGENDRE POLYNOMIALS AND OTHER SPECIAL FUNCTIONS

In this section we examine a number of sets of orthonormal functions on various intervals. This could be a mammoth undertaking.

Orthogonal Functions

The relations among such functions are tabulated in larger treatises, and the proofs of these relations are seldom of importance in quantum chemistry. Therefore, this section is devoted to a detailed examination of one of the special functions, the Legendre polynomials, and a cursory tabulation of others, including Laguerre and Hermite functions.

The Legendre polynomials are of importance in a number of quantum-chemical problems. They are the basis for the wave functions for angular momentum, and therefore occur in problems involving spherical motion, such as that of the electron in a hydrogen atom or the rotations of a molecule. In this context, the Legendre polynomials are important in describing the angular dependence of one-electron atomic orbitals; this dependence, in turn, forms the basis of the geometry of chemical compounds.

There are a number of ways of forming these polynomials, three of which are quite general.

(a) Schmidt orthogonalization of a linearly independent complete set.
(b) Solution of a differential equation.
(c) Use of a generating function.

In addition, we shall describe a formula for the Legendre polynomials. We begin with the orthogonalization procedure.

Legendre polynomials by Schmidt orthogonalization. The Legendre polynomials result from applying the Schmidt procedure to the set $\{x^n\}$ on the interval $[-1, 1]$. The choice of the interval $[-1, 1]$ is specific to the Legendre polynomials. We begin by letting ϕ_0 be proportional to $x^0 = 1$, and normalized on the interval:

$$N_0 = \int_{-1}^{1} 1 \, dx = 2 \qquad (2\text{-}72)$$

$$\phi_0 = (\tfrac{1}{2})^{1/2} \qquad (2\text{-}73)$$

Then, using Eqs. 2-56 and 2-57,

$$N_1 = \langle f_1 | f_1 \rangle - |\langle \phi_0 | f_1 \rangle|^2 = \langle f_1 | f_1 \rangle = \int_{-1}^{1} x^2 \, dx = \tfrac{2}{3} \qquad (2\text{-}74)$$

$$\phi_1 = N_1^{-1/2} (f_1 - \langle \phi_0 | f_1 \rangle \phi_0) = (\tfrac{3}{2})^{1/2} x \qquad (2\text{-}75)$$

Here, by realizing the oddness of the integrand, we have set $\langle \phi_0 | f_1 \rangle = 0$; and similarly, for all inner products where the sum of the indices is odd. This behavior was also noticed for the Hermite polynomials in the previous section.

Proceeding, we find

$$N_2 = \langle f_2 | f_2 \rangle - |\langle \phi_0 | f_2 \rangle|^2 \qquad (2\text{-}76)$$

$$\phi_2 = N_2^{-1/2}(f_2 - \langle \phi_0 | f_2 \rangle \phi_0) \qquad (2\text{-}77)$$

so we need to calculate

$$\langle f_2 | f_2 \rangle = \int_{-1}^{1} x^4 \, dx = \tfrac{2}{5} \qquad (2\text{-}78)$$

$$\langle \phi_0 | f_2 \rangle = \int_{-1}^{1} (\tfrac{1}{2})^{1/2} x^2 \, dx = (\tfrac{1}{2})^{1/2} \tfrac{2}{3} \qquad (2\text{-}79)$$

whence

$$N_2 = \tfrac{2}{5} - \tfrac{1}{2} \cdot \tfrac{4}{9} = \tfrac{8}{45} \qquad (2\text{-}80)$$

$$\phi_2 = (\tfrac{45}{8})^{1/2}(x^2 - \tfrac{1}{3}) = (\tfrac{5}{2})^{1/2}(\tfrac{3}{2}x^2 - \tfrac{1}{2}) \qquad (2\text{-}81)$$

The functions that are appearing are

$$\phi_0 = (\tfrac{1}{2})^{1/2} \cdot 1$$
$$\phi_1 = (\tfrac{3}{2})^{1/2} \cdot x$$
$$\phi_2 = (\tfrac{5}{2})^{1/2} \cdot (\tfrac{3}{2}x^2 - \tfrac{1}{2})$$
$$\phi_3 = (\tfrac{7}{2})^{1/2} \cdot (\tfrac{5}{2}x^3 - \tfrac{3}{2}x) \qquad (2\text{-}82)$$

The polynomials at the right are the Legendre polynomials; the whole orthonormal set of functions is $\{[(2n + 1)/2]^{1/2} P_n(x)\}$, where $P_n(x)$ signifies the *Legendre polynomial of rank n*.

Solution of Legendre's differential equation. The Legendre polynomials are the solutions of the differential equation

$$\frac{d}{dx}\left[(1 - x^2)\frac{df}{dx}\right] + l(l + 1)f = 0 \qquad (2\text{-}83)$$

where l is a positive integer or zero. That the polynomials that have been generated by orthogonalization of $\{x^n\}$ on $[-1, 1]$ are solutions of Eq. 2-83 can be verified by direct substitution. However, we may derive this result directly using a method that is a prototype for the solution of many differential equations.

We attempt a power series solution for f, $f = \sum_n a_n x^n$. Substituting this trial solution (often called *Ansatz*, from the German) into Eq. 2-83, we obtain

$$\frac{d}{dx}\left[(1-x^2)\sum_n na_n x^{n-1}\right] + \left[l(l+1)\sum_n a_n x^n\right] = 0$$

$$\frac{d}{dx}\left[\sum_n na_n x^{n-1} - \sum_n na_n x^{n+1}\right] + \sum_n l(l+1)a_n x^n = 0$$

$$\sum_n n(n-1)a_n x^{n-2} - \sum_n n(n+1)a_n x^n + \sum_n l(l+1)a_n x^n = 0$$
(2-84)

If we collect all similar powers of x, we obtain

$$\sum_n x^n [(n+2)(n+1)a_{n+2} - n(n+1)a_n + l(l+1)a_n] = 0$$
(2-85)

but, for the series to be identically zero, each term must be identically zero, since the functions $\{x^n\}$ are linearly independent. This gives the *recursion relation* for the expansion coefficients,

$$a_{n+2} = \frac{[n(n+1) - l(l+1)]a_n}{(n+2)(n+1)}$$
(2-86)

Equation 2-86 gives a rule for finding every other coefficient starting with the first. That is, if $a_0 = 1$, then Eq. 2-86 gives $a_2 = -l(l+1)/2$, $a_4 = [6 - l(l+1)]/12 \cdot [-l(l+1)/2]$, and so forth. The coefficients depend on the value of l. Suppose we continue on, finding all the even coefficients. If the *rank* of the equation, l, is even, eventually n will equal l for some term, $n(n+1)$ will equal $l(l+1)$, and the following coefficient, a_{n+2}, will equal zero. Then all following coefficients will also equal zero. In other words, if l is even, the even terms of the power series cut off at $n = l$. The odd terms, of course, keep on going. This infinite series of odd powers diverges at $x = \pm 1$, and is not of physical interest; the finite series of even powers gives

$$l = 0 \quad a_0 = 1 \quad f_0 = 1$$
$$l = 2 \quad a_0 = 1$$
$$\quad\quad\quad a_2 = -3 \quad f_2 = -3x^2 + 1$$

and so forth, which are proportional to the Legendre polynomials derived by orthogonalization. It has become customary to choose a_0 such that the polynomial $f_n = 1$ at $x = +1$. Then

$$l = 0 \quad a_0 = 1 \quad\quad P_0 = 1$$
$$l = 2 \quad a_0 = -\tfrac{1}{2}$$
$$\quad\quad\quad a_2 = \tfrac{3}{2} \quad\quad P_2 = \tfrac{3}{2}x^2 - \tfrac{1}{2}$$

and so on.

Similarly, if l is odd, the series in odd powers terminates at $n = l$, and the series in even powers is an infinite series, divergent at $x = \pm 1$. Again, we choose a_0 such that $f_n(+1) = 1$. Then

$$l = 1 \quad a_1 = 1 \quad\quad P_1 = x$$
$$l = 3 \quad a_1 = -\tfrac{3}{2}$$
$$\quad\quad\quad a_3 = \tfrac{5}{2} \quad\quad P_3 = \tfrac{5}{2}x^3 - \tfrac{3}{2}x$$

and so forth. The method of power series substitution is very powerful. The usual result is a recursion relation which may be interpreted, within the context of the problem, as we have done here.

The generating function for Legendre polynomials. A generating function for a set of functions is some function of two variables such that when this function is expanded as a power series in one of the variables, the coefficients are the set of functions in the other variable. The generating function for Legendre polynomials is

$$F(x, t) = (1 - 2xt + t^2)^{-1/2} = \sum_n P_n(x)t^n \qquad (2\text{-}87)$$

If the generating function is written as $[1 - (2xt - t^2)]^{-1/2}$, it may be expanded in the binomial series

$$F(x, t) = 1 + \tfrac{1}{2} \cdot \tfrac{1}{1!}(2xt - t^2) + \tfrac{1}{2} \cdot \tfrac{3}{2} \cdot \tfrac{1}{2!}(2xt - t^2)^2$$
$$+ \tfrac{1}{2} \cdot \tfrac{3}{2} \cdot \tfrac{5}{2} \cdot \tfrac{1}{3!}(2xt - t^2)^3 + \cdots \qquad (2\text{-}88)$$

and then rearranged in powers of t:

$$F(x, t) = 1 + (x)(t) + \left(\frac{-1}{2} + \frac{3}{2}x^2\right)(t^2)$$

Orthogonal Functions

$$+ \left(-\frac{3}{2}x + \frac{5}{2}x^3\right)(t^3) + \cdots$$

$$= \sum_n P_n(x) t^n \qquad (2\text{-}89)$$

Often many useful relations may be derived from a generating function; however, there are no general methods for constructing a generating function.

Rodrigues' formula for the Legendre polynomials. As a final example of a formula for finding Legendre polynomials, we derive Rodrigues' formula,

$$P_l(x) = \frac{1}{2^l l!} \frac{d^l}{dx^l}(x^2 - 1)^l \qquad (2\text{-}90)$$

Consider the differential equation

$$(1 - x^2)\frac{dy}{dx} + 2lxy = 0 \qquad (2\text{-}91)$$

which can be solved by the elementary "separation of variables" technique, giving

$$\frac{dy}{y} = \frac{-2lx\,dx}{1 - x^2} \qquad (2\text{-}92)$$

whose integral is

$$y = (1 - x^2)^l \qquad (2\text{-}93)$$

If the original differential equation, Eq. 2-91, is now differentiated $(l + 1)$ times,[4] the result is

$$(1 - x^2)\frac{d^{l+2}y}{dx^{l+2}} + (l + 1)(-2x)\frac{d^{l+1}y}{dx^{l+1}}$$

$$+ \frac{l(l + 1)}{2}(-2)\frac{d^l y}{dx^l} + 2xl\frac{d^{l+1}y}{dx^{l+1}}$$

$$+ 2l(l + 1)\frac{d^l y}{dx^l} = 0 \qquad (2\text{-}94)$$

---

[4] In this derivation, use has been made of the expansion of the *n*th derivative of a product: $d^n/dx^n(uv) = (d^n u/dx^n)v + n(d^{n-1}u/dx^{n-1})(dv/dx) + \cdots$, a series with binomial coefficients.

If we now substitute into this equation $f = d^l y/dx^l$, we find that f is a solution of Legendre's differential equation:

$$(1 - x^2)\frac{d^2 f}{dx^2} - 2x\frac{df}{dx} + l(l+1)f = 0$$

$$\frac{d}{dx}\left[(1 - x^2)\frac{df}{dx}\right] + l(l+1)f = 0 \qquad (2\text{-}95)$$

Therefore, $f = d^l y/dx^l = d^l/dx^l(1 - x^2)^l$ is a solution of Legendre's equation. However, the solutions of Legendre's equation are the Legendre polynomials. We have only to insure $f(+1) = 1$, as we required, to establish the relation desired. If the formula is expanded as a derivative of a product,[4] we obtain

$$f = \frac{d^l}{dx^l}[(x+1)^l(x-1)^l]$$

$$= [(x-1)^l]\left[\frac{d^l}{dx^l}(x+1)^l\right]$$

$$+ l\left[\frac{d}{dx}(x-1)^l\right]\left[\frac{d^{l-1}}{dx^{l-1}}(x+1)^l\right] + \cdots$$

$$+ l\left[\frac{d^{l-1}}{dx^{l-1}}(x-1)^l\right]\left[\frac{d}{dx}(x+1)^l\right]$$

$$+ \left[\frac{d^l}{dx^l}(x-1)^l\right][(x+1)^l] \qquad (2\text{-}96)$$

Evaluated at $x = 1$, all the terms but the last are zero, since each will contain powers of $(x - 1)$. The last term, however, does not contain powers of $(x - 1)$: $(d^l/dx^l)(x - 1)^l = l!$ and $(x + 1)^l = 2^l$. Therefore, $f(1) = 2^l l!$; to make $f(1) = 1$, and agree with the Legendre polynomials, we divide by the factor

$$P_l(x) = \frac{1}{2^l l!}\frac{d^l}{dx^l}(x^2 - 1)^l \qquad (2\text{-}97)$$

Having described four different methods for finding Legendre polynomials (and there are others, too), we shall now use these formulas to derive some of the salient properties of the polynomials.

Evaluation of $P_l(x)$ at $x = -1$, $x = 0$. We have restricted our definition of the polynomials by the boundary condition $P_l(+1) =$

+1, but it is also useful to know the values of $P_l(x)$ at the other boundary, $P_l(-1)$, and in the middle of the interval, $P_l(0)$. Rodrigues' formula is useful for obtaining $P_l(-1)$. From an expansion of $P_l(x) = (1/2^l l!)(d^l/dx^l)(1-x^2)^l$ similar to that given in Eq. 2-96, it is clear that only the first term will be nonzero at $x = -1$, since all other terms involve powers of $(x+1)$. That first term at $x = -1$ is $(-2)^l l!$. Then,

$$P_l(-1) = \frac{1}{2^l l!}(-2)^l l! = (-1)^l \qquad (2\text{-}98)$$

That is, $P_l(-1)$ is $+1$ or -1 as l is even or odd.

We may also use Rodrigues' formula to evaluate $P_l(0)$. The binomial expansion of $(x^2-1)^l$ gives

$$(x^2-1)^n = x^{2n} + \frac{n(-1)}{1!}x^{2n-2} + \cdots$$
$$+ \frac{(-1)^k n(n-1)(n-2)\cdots(n-k+1)}{k!}x^{2n-2k}$$
$$+ \cdots + n(-1)^{n-1}x^2 + (-1)^n \qquad (2\text{-}99)$$

For the polynomial P_n, we require the nth derivative of Eq. 2-99. All terms in the binomial expansion afford zero for the result of taking n derivatives and evaluating at $x = 0$, except a possible term in x^n. If there is a term in x^n, the nth derivative, at $x = 0$, will give $n!$. There will be a term in x^n only if n is even, so that there can be some value of k in Eq. 2-99 for which $2n - 2k = n$, namely $k = n/2$. From these considerations, we conclude

$$\frac{d^n}{dx^n}(x^2-1)^n \bigg|_{x=0} = 0 \qquad (2\text{-}100a)$$

if n is odd, and

$$\frac{d^n}{dx^n}(x^2-1)^n \bigg|_{x=0} = (-1)^{n/2}\frac{n(n-1)\cdots\left(n-\frac{n}{2}+1\right)n!}{(n/2)!} \qquad (2\text{-}100b)$$

if n is even.

The nonzero result can be simplified. Let $2l = n$, and then

$$P_{2l}(0) = \left(\frac{1}{2^{2l}(2l)!}\right)(-1)^l \frac{(2l)(2l+1)\cdots(l+2)(l+1)(2l)!}{l!}$$

$$= \left(\frac{1}{2^{2l}}\right)(-1)^l$$
$$\times \frac{(2l)(2l-1)\cdots(l+2)(l+1)(l)(l-1)\cdots(2)(1)}{l!l!}$$
$$= \left(\frac{1}{2^{2l}}\right)(-1)^l \frac{(2l-1)(2l-3)\cdots(3)(1)(2^l)}{l!}$$
$$= (-1)^l \frac{(2l-1)!!}{2^l l!} \tag{2-101}$$

where the symbol $(2l-1)!!$, read "$2l-1$ double factorial," indicates a product of every other integer, $n!! = n(n-2)(n-4)\cdots$. This result may also be derived using the generating function for the Legendre polynomials. For, setting $x=0$ in $F(x,t) = [1 - 2xt + t^2]^{-1/2}$, we obtain

$$F(0,t) = \sum_n P_n(0)t^n = (1+t^2)^{-1/2}$$
$$= 1 - \frac{1}{1!}\cdot\frac{1}{2}t^2 + \frac{1}{2!}\cdot\frac{1}{2}\cdot\frac{3}{2}t^4 - \cdots$$
$$= \sum_{n \text{ even}} \frac{(n-1)!!(-1)^{n/2}}{(2^{n/2})}\left(\frac{n}{2}\right)! t^n$$
$$= \sum_l \frac{(2l-1)!!(-1)^l}{2^l l! t^{2l}} \tag{2-102}$$

A plot of the first four Legendre polynomials is shown in Fig. 2–6. The values of $P_l(\pm 1)$ and $P_l(0)$ are apparent.

Recursion relations among Legendre polynomials. As a further example of the use of the generating function, we derive two recursion relations interrelating the Legendre polynomials. By differentiating Eq. 2–87 with respect to t, we obtain

$$\frac{\partial F(x,t)}{\partial t} = (x-t)(1-2xt+t^2)^{-3/2} = \sum_n nt^{n-1}P_n(x) \tag{2-103}$$

which may be rearranged to give

$$(x-t)(1-2xt+t^2)^{-1/2} = (1-2xt+t^2)\sum_n nt^{n-1}P_n(x) \tag{2-104}$$

Orthogonal Functions

$$(x - t) \sum_n t^n P_n(x) = (1 - 2xt + t^2) \sum_n nt^{n-1} P_n(x)$$

Since the powers of t are linearly independent, we can equate coefficients of t^n:

$$xP_n - P_{n-1} = (n + 1)P_{n+1} - 2nxP_n + (n - 1)P_{n-1}$$

or

$$(n + 1)P_{n+1} - (2n + 1)xP_n + nP_{n-1} = 0 \qquad (2\text{–}105)$$

which interrelates the various Legendre polynomials. For instance, given P_1 and P_2, we may find P_3 by using $n = 2$ in Eq. 2-105:

$$\begin{aligned}
P_3 &= \frac{(2 \cdot 2 + 1)xP_2 - 2P_1}{2 + 1} = \frac{3P_2 - 2P_1}{3} \\
&= \frac{5x(\frac{3}{2}x^2 - \frac{1}{2}) - 2x}{3} = \frac{5}{2}x^2 - \frac{5}{6}x - \frac{2}{3}x \\
&= \frac{5}{2}x^3 - \frac{3}{2}x \qquad (2\text{–}106)
\end{aligned}$$

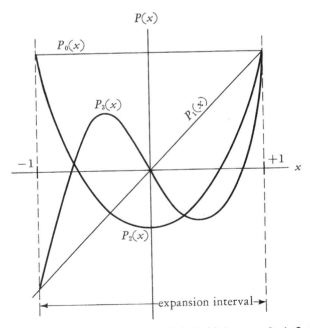

Figure 2-6 Legendre polynomials $P_n(x)$ for $n = 0, 1, 2, 3$.

This recursion relation can be used to generate higher Legendre polynomials.

Another relation, obtained by differentiating Eq. 2-87 with respect to x, is

$$\frac{\partial F(x, t)}{\partial x} = t(1 - 2xt + t^2)^{-3/2} = \sum_n P'_n(x)t^n \qquad (2\text{-}107)$$

which yields

$$t \sum_n P_n t^n = (1 - 2xt + t^2) P'_n t^n \qquad (2\text{-}108)$$

or, for t^n,

$$P_{n-1} = P'_n - 2xP'_{n-1} + P'_{n-2} \qquad (2\text{-}109)$$

This recursion relation relates the Legendre polynomials to their first derivatives. Many other useful relations may be derived by analogous manipulations of the generating function.

Spherical harmonics: an orthonormal set in two variables. Legendre's differential equation, Eq. 2-83, is a special case of the equation

$$\frac{d}{dx}\left[(1 - x^2)\frac{df}{dx}\right] + \left[l(l + 1) - \frac{m^2}{1 - x^2}\right]f = 0 \qquad (2\text{-}110)$$

where m is an integer. If $m = 0$, Eq. 2-110 is identical to Eq. 2-83. The solutions to this equation are known as the associated Legendre functions, $P_l{}^m(x) = (-1)^m(1 - x^2)^{m/2} d^m P_l(x)/dx^m$. The associated Legendre functions are marked by two indices, m and l. The associated Legendre functions are also orthogonal on the interval $[-1, 1]$ in the sense $\langle P_{l'}{}^m | P_l{}^m \rangle = 0$. The inner product of two associated Legendre functions for different l but the same m is zero. The norm of the associated Legendre functions is

$$N(P_l{}^m) = (2)(l + m)!/(2l + 1)(l - m)!$$

The set of functions

$$\left\{ \sqrt{\frac{(2l + 1)(l - m)!}{2(l + m)!}} \, P_l{}^m(x) \right\}$$

defined on $[-1, 1]$, for a fixed value of m and $l = m, m + 1, \ldots$ are complete and orthonormal. The Legendre polynomials themselves are an example of this set, for, if $m = 0$, this set is identical to the set formed by Schmidt orthogonalization.

The associated Legendre functions may be rewritten as functions of an angle variable θ, on $0 \leq \theta \leq \pi$, by replacing x by $\cos \theta$. The functions $\{P_l^m(\theta)\}$ are therefore a complete orthogonal set of functions on the interval $[0, \pi]$ in the variable θ. However, we saw in Section 2-3 that the set $\{(2\pi)^{-1/2} e^{im\phi}\}$ is a complete orthonormal set of the interval $[0, 2\pi]$ in the variable ϕ. By multiplying each of the functions $[(2l + 1)(l - m)!/2(l + m)!]^{1/2} P_l^m (\cos \theta)$ by the members of this other complete, orthonormal set, $\{(2\pi)^{-1/2} e^{im\phi}\}$, we obtain a set of functions, complete and orthonormal in two variables, called the *spherical harmonics*:

$$Y_{l,m}(\theta, \phi) = \left[\frac{(2l + 1)(l - m)!}{4\pi(l + m)!} \right]^{1/2} P_l^{|m|}(\cos \theta) e^{im\phi} \qquad (2\text{-}111)$$

with

$$Y_{l,-m}(\theta, \phi) = (-1)^m Y_{l,m}^* \qquad (2\text{-}112)$$

The index m is usually called the *order*, and the index l the *rank* of the spherical harmonics. The rank assumes the values $l = 0, 1, 2, \ldots$; the order m, the values $-l, (-l + 1), \ldots, 0, \ldots, (l - 1), l$.

We have now spent some time discussing the ins and outs of Legendre polynomials and the associated Legendre functions. Lest the student miss the forest for the trees, we now recapitulate the salient points of our discussion.

1. A set of polynomials can be formed by Schmidt orthogonalization from the complete set $\{x^n\}$ on $[-1, 1]$. Except for a normalization factor, these are the Legendre polynomials.

2. These same polynomials are solutions of the differential equation $(d/dx)[(1 - x^2) df/dx] + l(l + 1)f = 0$.

3. These same polynomials are the coefficients in a series expansion in powers of t of $F(x, t) = (1 - 2xt + t^2)^{-1/2}$.

4. These same polynomials are given by $(1/2^l l!)(d^l/dx^l)(1 - x^2)^l$.

5. A number of relations may be derived from points 3 and 4.

6. Functions $P_l^m(x) = (-1)^m (1 - x^2)^{m/2} (d^m P_l/dx^m)$, called the associated Legendre functions, are solutions of the differential equation $(d/dx)[(1 - x^2)(df/dx)] + [l(l + 1) - (m^2/(1 - x^2))]f = 0$. These functions are orthogonal for different l and the same m.

7. The functions of two variables (the angles θ and ϕ), $Y_{l,m}(\theta, \phi)$, formed by multiplying the normalized associated Legendre functions by the harmonic functions $e^{im\phi}/(2\pi)^{-1/2}$, are orthonormal and complete on the intervals $0 \leq \theta \leq \pi$ and $0 \leq \phi \leq 2\pi$.

This section has attempted to present the two following ideas.

1. An illustration of the ways of forming the important special functions of mathematical physics: orthogonalization, power series solution of a differential equation, and generating function.

2. A record for the student's use—but by no means his memorization—of some of the important properties of a number of functions which occur in chemistry. The behavior of these functions, even in a qualitative way, gives useful information about chemical systems. For example, since $P_{2l+1}(0) = 0$, the p and f atomic orbitals have no electron density perpendicular to their axes. The student should endeavor to become as familiar with the Legendre (and other) functions as he is with sines and cosines, so that he may easily picture the behavior of chemical systems which these functions describe.

For the most part, these and other facts can be found elsewhere. Some students, hopefully, will come to enjoy the mathematical manipulations—investigating the properties of integrals and derivatives—but this is not usually the most important part of attacking a physical problem.

This section concludes with a summary of the occurrence of the special functions of use in quantum chemistry.

Special functions of quantum chemistry

Harmonic, $\{e^{imx}\}$.
    Differential equation: $(d^2f/dx^2) + m^2f = 0$.
    Orthogonalization: on $[0, 2\pi]$.
    Normalization: multiply by $(2\pi)^{-1/2}$.
    Occurrence: translational motion.

Legendre, $\{P_l(x)\}$.
Differential equation: $(1 - x^2)(d^2f/dx^2) - 2x\,df/dx + l(l+1)f = 0$.
Orthogonalization: on $[-1, 1]$ from $\{x^n\}$.
Normalization: $[(2l+1)/2]^{1/2}$.
Generating function: $(1 - 2xt + t^2)^{-1/2} = \sum_n P_n(x)t^n$.

Occurrence: angular motion.

Associated Legendre, $\{P_l^m(x)\}$.
Differential equation: $(1 - x^2)(d^2f/dx^2) - 2x(df/dx) + [l(l+1) - m^2/(1-x^2)]f = 0$.
Orthogonalization: for same m, different l, on $[-1, 1]$.
Normalization: $[(2l+1)(l-m)!/2(l+m)!]^{1/2}$.
Occurrence: angular motion.

Laguerre, $\{L_n(x)\}$.
Differential equation:
$x(d^2f/dx^2) + (df/dx) - (\tfrac{1}{2} + x/4 + n)f = 0$.
Orthogonalization: on $[0, \infty)$, from $\{e^{-x/2}x^n\}$.
Normalization: $1/n!$.
Generating function: $\exp[-xt/(1-t)]/(1-t) = \sum_n L_n(x)t^n$.

Occurrence: radial motion.

Hermite, $\{H_n(x)\}$.
Differential equation: $(d^2f/dx^2) - 2x(df/dx) + 2nf = 0$.
Orthogonalization: on $(-\infty, \infty)$, from $\{\exp[-x^2/2]x^n\}$.
Normalization: $[1/2^n n! \pi^{1/2}]^{1/2}$.
Generating function: $\exp[2tx - t^2] = \sum_n H_n(x)t^n/n!$.

Occurrence: harmonic oscillator.

Problems

1. Show that the set of functions $\{\sin mx, \cos nx\}$, for positive, integral values of m and n, are orthogonal on the interval $[-\pi, \pi]$.

2. Classify the following functions as even, odd, or neither even nor odd. If the function is neither even nor odd, decompose the function into the sum of an even function and an odd function. (a) x^2; (b) x^3; (c) $x \sin x$; (d) $x^3 \cos nx$ (n integral); (e) x^4; (f) $\log[(1+x)/(1-x)]$; (g) e^x; (h) e^{ix}.

3. Show that the sets of functions $\{\sin mx\}$, $\{\cos mx\}$ are orthogonal on

the interval $[0, \pi]$. Find, on this interval, the norms of these functions. Calculate the expansion coefficients for an expansion in these sets of functions.

4. Expand the function $\pi x - x^2$ on the interval $[-\pi, \pi]$; on the interval $[0, \pi]$ in sines; on the interval $[0, \pi]$ in cosines. Which expansion converges most rapidly on $[0, \pi]$?

5. Expand the function $\sin^2 x$ in a sine series and a cosine series on $[0, \pi]$. Which expansion converges most rapidly? Compare the situation with that of Problem 4.

6. Using the information and results of Problems 4 and 5 and the expansion theorem, calculate the inner product $\langle \sin^2 x \mid \pi x - x^2 \rangle$ on $[0, \pi]$.

7. Show that the functions $\{e^{imx}\}$ are orthogonal on the interval $[0, 2\pi]$. Find the norm of these functions. Calculate the expansion coefficients for an expansion in these functions. Compare these coefficients with the usual expansion coefficients in a Fourier series.

8. A *square wave* is defined by $f(x) = a$ for $0 < x < \pi$, $f(x) = -a$ for $-\pi < x < 0$. Carry out a Fourier series expansion for the square wave.

9. A *triangular wave* is defined by $f(x) = x + \pi$ on $-\pi \leq x \leq 0$, $f(x) = \pi - x$ on $0 \leq x \leq \pi$. Carry out a Fourier series expansion for the triangular wave.

10. Show that the functions $\{\exp[2ni\pi x/(b-a)]\}$ are orthogonal on the interval $[a, b]$.

11. A Taylor series about zero is an expansion in powers of x. Is the Taylor series an expansion in orthogonal functions? Show how different sets of orthonormal functions may be constructed from the powers of x used in the Taylor series.

12. The functions $\{\sin^n x, \cos^n x\}$, for $n = 0, 1, 2, \ldots$ are not an orthonormal set. Show that the orthonormal set which can be constructed from this set on the interval $[-\pi, \pi]$ are just the functions of Problem 1.

13. Use the differential equation, Eq. 2–110, to prove that the associated Legendre functions are, in fact, orthogonal as stated.

14. Verify the orthogonalization of the Laguerre polynomials as stated in the list of special functions; verify their normalization.

15. Verify the normalization of the Hermite polynomials as stated in the list of special functions.

16. Expand the function e^x in Legendre, Hermite, and Laguerre polynomials, and compare the results. Notice that the expansion intervals are quite different.

17. The discussion of the Schmidt procedure was somewhat arbitrary in the choice of the expansion interval. With a number of trial examples, show

Orthogonal Functions

what effect changing the expansion interval has on the orthogonal functions.

18. In spectroscopy, *selection rules* govern what transitions between energy levels are allowed. For so-called *electric dipole transitions*, the intensity of a spectroscopic absorption is proportional to $|\langle \psi_1 | x\psi_2 \rangle|^2$, which is called the *transition dipole moment*. Calculate the transition dipole moment for the following transitions of the harmonic oscillator, where $\psi_n(x) = \exp[-x^2/2]H_n(x)$.

$$\psi_1 \to \psi_2 \qquad \psi_2 \to \psi_3$$
$$\psi_1 \to \psi_3 \qquad \psi_3 \to \psi_4$$
$$\psi_1 \to \psi_4$$

Do you see a pattern emerging? Can you formulate and prove the general selection rules for the harmonic oscillator?

3

Linear Algebra

We mentioned briefly in Chapter 1 that quantum mechanics may be looked upon from two points of view. The first, Schrödinger's wave mechanics, casts eigenvalue equations in the form of differential equations of functions of one or more variables. With this in mind, we studied in depth the concept of orthogonal functions in Chapter 2 in order to establish the kind of language needed to discuss the Schrödinger wave functions. We concluded by solving one such differential equation, Legendre's differential equation, and related the solution to the established concepts and behavior of orthonormal functions.

The second point of view from which quantum mechanics can be studied is Heisenberg's matrix mechanics. In this chapter we shall be concerned with building up the vocabulary of algebra that is used in studying matrix mechanics. We shall also introduce in a more complete fashion the entities called operators, which were mentioned in Chapter 1, and derive some traditional results of interest and importance in quantum chemistry.

3-1 INTRODUCTION

With no further ado, we begin with a barrage of definitions.

Definition An n-dimensional *vector* α is an n-tuple, or list of n numbers: $\alpha = (\alpha_1, \alpha_2, \ldots, \alpha_n)$. The n numbers $\alpha_1, \alpha_2, \ldots, \alpha_n$ are

Linear Algebra

called the *components* of the vector; they are arranged in a prespecified order; they may be real or complex.

We shall denote vectors by Greek letters. An example is a conventional physical vector in three-dimensional space represented by three numbers that denote the x, y, and z components, respectively.

> **Definition** A *vector space* is a set of vectors with two properties:
> (a) A unique, commutative sum is defined by $\alpha^1 + \alpha^2 = \alpha^2 + \alpha^1 = (\alpha_1^1 + \alpha_1^2, \alpha_2^1 + \alpha_2^2, \ldots, \alpha_n^1 + \alpha_n^2)$.
> (b) Any vector α and any scalar c define a unique product $c\alpha = (c\alpha_1, c\alpha_2, \ldots, c\alpha_n)$ which is both distributive [i.e., $c(\alpha + \beta) = c\alpha + c\beta$] and associative [i.e., $(c_1 c_2)\alpha = c_1(c_2 \alpha)$].

The student should realize that these properties are familiar in the case of three-dimensional physical vectors. We denote members of a *set* of vectors with a superscript, $\{\alpha^i\}$, but components of a single vector with a subscript.

> **Definition** A *Euclidean vector space* is a vector space for which the components of the vectors are real, and such that with any two vectors we may associate a real *inner product* $\langle \alpha \mid \beta \rangle = \alpha_1 \beta_1 + \alpha_2 \beta_2 + \cdots + \alpha_n \beta_n$, that is symmetric [$\langle \alpha \mid \beta \rangle = \langle \beta \mid \alpha \rangle$], bilinear [$\langle \alpha + \beta \mid \gamma \rangle = \langle \alpha \mid \gamma \rangle + \langle \beta \mid \gamma \rangle$], and positive [$\langle \alpha \mid \alpha \rangle \geq 0$].

This definition of inner product uses the same symbol, $\langle \mid \rangle$, that was used for inner products of functions.

A slightly different, and more general, definition of inner product is involved in the definition of a slightly different, and more general, vector space.

> **Definition** A *Hermitian vector space* is a vector space for which the components of the vectors may be complex, and such that with any two vectors α and β, we may associate a complex inner product $\langle \alpha \mid \beta \rangle = \alpha_1^* \beta_1 + \alpha_2^* \beta_2 + \cdots + \alpha_n^* \beta_n$, which is Hermitian [$\langle \alpha \mid \beta \rangle = \langle \beta \mid \alpha \rangle^*$], bilinear [$\langle \alpha + \beta \mid \gamma \rangle = \langle \alpha \mid \gamma \rangle + \langle \beta \mid \gamma \rangle$], and positive [$\langle \alpha \mid \alpha \rangle \geq 0$].

Two comparisons should be made. First, compare the definition of inner product that appears here

$$\langle \alpha \mid \beta \rangle = \sum_i \alpha_i^* \beta_i \qquad (3\text{-}1)$$

with that of Chapter 2 (Eq. 2-1)

$$\langle f \mid g \rangle = \int f(x)^* g(x)\, dx$$

The difference is that the vector inner product is defined by a sum over a discrete index, whereas the function inner product is defined by an integral over a continuous variable. This difference—the exchange of a discrete index for a continuous variable—is much of what distinguishes the Heisenberg from the Schrödinger picture.

Again, notice the difference between Euclidean and Hermitian vector spaces: the inner product in a Euclidean vector space is real and symmetric; in a Hermitian vector space the inner product is complex and Hermitian. This property of inner products in Hermitian vector spaces should recall the analogous property of the inner product of two functions: *Transposing an inner product gives the complex conjugate of that inner product.*

> **Definition** Two vectors are said to be *orthogonal* if their inner product is zero.

This definition is strictly parallel to the definition of orthogonality of functions; it makes the concept of orthogonality one of sweeping generality, and establishes a connection to the familiar property of perpendicular vectors.

> **Definition** Again, by analogy with the material on functions which has gone before, we define the *norm* of a vector by $N(\alpha) = \langle \alpha \mid \alpha \rangle = |\alpha|^2$.

The norm corresponds to the length of a vector in physical three-dimensional space. Two final definitions continue the parallel structure of the algebra of vector spaces with the calculus of functions.

> **Definition** A vector is said to be *normalized* if its norm is one.

> **Definition** An *orthonormal set of vectors* is a set of vectors, each of which is normalized, and each of which is orthogonal to every other vector in the set.

As an illustration of these definitions, we prove Schwarz's inequality.

> **Theorem** *Schwarz's inequality:* $|\langle \alpha \mid \beta \rangle| \leq |\alpha| \cdot |\beta|$

Linear Algebra

If either α or β is zero, the equality sign obtains, and the theorem is trivial. If neither α nor β is zero, we construct the theorem with the aid of two arbitrary scalars, c and d. By the positiveness property of inner products,

$$0 \leq \langle c\alpha + d\beta \mid c\alpha + d\beta \rangle \tag{3-2}$$

and, by bilinearity,

$$\begin{aligned} 0 &\leq c^* \langle \alpha \mid c\alpha + d\beta \rangle + d^* \langle \beta \mid c\alpha + d\beta \rangle \\ &\leq c^* c \langle \alpha \mid \alpha \rangle + c^* d \langle \alpha \mid \beta \rangle + d^* c \langle \beta \mid \alpha \rangle + d^* d \langle \beta \mid \beta \rangle \end{aligned} \tag{3-3}$$

This equation is true for any value of c and d; therefore, it must be true for the particular values $c = -\langle \alpha \mid \beta \rangle$, and $d = \langle \alpha \mid \alpha \rangle$. With this, Eq. 3-3 yields

$$\begin{aligned} 0 &\leq c^* cd - c^* dc - d^* cc^* + d^* d \langle \beta \mid \beta \rangle \\ &\leq d[-c^* c + d^* \langle \beta \mid \beta \rangle] \end{aligned} \tag{3-4}$$

so that

$$0 \leq -|\langle \alpha \mid \beta \rangle|^2 + \langle \alpha \mid \alpha \rangle \langle \beta \mid \beta \rangle \tag{3-5}$$

or

$$|\langle \alpha \mid \beta \rangle|^2 \leq |\alpha|^2 |\beta|^2 \tag{3-6}$$
$$|\langle \alpha \mid \beta \rangle| \leq |\alpha||\beta| \tag{3-7}$$

Having established definitions of inner product and norm which apply to both functions and vectors, we may apply to vectors results which were studied in detail for functions. We proceed to a consideration of sets of vectors; our results will, to a large extent, parallel the results obtained in our earlier study of sets of functions.

The concept of completeness of a set of functions has an analog in terms of vector spaces.

Definition A vector space is *spanned* by a set of vectors $\{\alpha^i\}$ if any vector in the vector space can be expressed as a linear combination of the set $\{\alpha^i\}$.

This definition should be easily pictured in terms of the usual three-dimensional Euclidean vector space (which we shall call 3D space hereafter). As an example, the vectors $(1, 0, 0)$, $(0, 1, 0)$, $(0, 0, 1)$ *span* 3D space. These are just the unit vectors in the three coordinate directions, and it is a familiar fact that any vector may be expressed as a linear combination of the three unit vectors. This definition prompts a more rigorous interpretation of the idea of dimension.

> **Definition** The *dimension* of a vector space is the minimum number of vectors required to span the vector space.

In the example given above, no two vectors are sufficient to span 3D space, but four vectors, say, $(1, 0, 0)$, $(0, 1, 0)$, $(0, 0, 1)$, and $(1, 1, 1)$ are redundant. Any one of them could be omitted and the remaining three would span 3D space. Of course, not all sets of three vectors span 3D space. The vectors $(1, 0, 0)$, $(0, 1, 0)$, $(1, 1, 0)$, although three (the dimension) in number, do not span 3D space. Hopefully, the reader can see why. These vectors are linearly dependent, and so one is redundant. The concept of linear independence thus makes an appearance again in our consideration of vector spaces.

> **Definition** A set of vectors $\alpha^1, \alpha^2, \ldots, \alpha^n$ is *linearly independent* if the equation
> $$c_1\alpha^1 + c_2\alpha^2 + \cdots + c_n\alpha^n = \sum_i c_i \alpha^i = 0 \qquad (3\text{-}8)$$

These properties of vector spaces allow the convenient concept of a basis.

> **Definition** A *basis* of (or for) a vector space is *some* linearly independent set of vectors that spans the vector space; an *orthonormal basis* of a vector space is some orthonormal set which spans the space.

These definitions should recall the process used to form an orthonormal set of functions from a linearly independent set, the Schmidt orthogonalization procedure. This procedure may be used for vectors with no modification. If the linearly independent set is notated $\{\alpha^i\}$ and the orthonormal set $\{\phi^i\}$, the pertinent formulas are

$$\phi^k = N_k^{-1/2}(\alpha^k - \sum_{j=0}^{k-1} \langle \phi^j \mid \alpha^k \rangle \phi^j) \qquad (3\text{-}9)$$

Linear Algebra

$$N_k = \langle \alpha^k \mid \alpha^k \rangle - \sum_{j=0}^{k-1} |\langle \phi^j \mid \alpha^k \rangle|^2 \qquad (3\text{-}10)$$

With this background, in almost all respects parallel to that of orthonormal functions, it should be clear that all the algebraic "machinery" is now available for the expansion of vectors in terms of an orthonormal set of vectors. Precisely the same derivation that was set aside from the text in Chapter 2, Eqs. 2-19 through 2-24, applies to the vector problem. We have, therefore, the expansion of an arbitrary vector ξ in the set $\{\phi^i\}$, according to

$$\xi = \sum_i c_i \phi^i \qquad (3\text{-}11)$$

with the expansion coefficients (perhaps complex)

$$c_i = \langle \phi^i \mid \xi \rangle \qquad (3\text{-}12)$$

We may also borrow from Chapter 2 the expansion theorem for inner products,

$$\langle \xi \mid \eta \rangle = \sum_i \langle \xi \mid \phi^i \rangle \langle \phi^i \mid \eta \rangle \qquad (3\text{-}13)$$

where the set $\{\phi^i\}$ is orthonormal.

One departure—a useful one—from the parallel study of vector spaces and functions is that for finite-dimensional vector spaces we may carry out straightforward tests to determine whether or not a set of vectors is linearly independent. The reader may recall that this question was generally bypassed offhand in talking about sets of functions, and we relied there more on intuition than on thorough reasoning.

For sets of vectors, however, we may carefully make an analysis to determine if the set (finite) is linearly independent. This analysis makes a fitting conclusion to this section and an introduction to the next, since the mathematical element called a *matrix* appears.

In considering linear independence, we want to try to find some c_i's, not zero, that satisfy Eq. 3-8. In order to make this search something better than trial and error, a simple procedure has been developed.

Line up the vectors in rows. Suppose there are three vectors α, β, γ, each of which has four components. Our lineup looks like this:

$$\begin{array}{cccc} \alpha_1 & \alpha_2 & \alpha_3 & \alpha_4 \\ \beta_1 & \beta_2 & \beta_3 & \beta_4 \\ \gamma_1 & \gamma_2 & \gamma_3 & \gamma_4 \end{array}$$

Begin with the bottom row. We can surely find some multiple of α_1, which, when added to γ_1, gives 0. In fact, that multiple is $-\gamma_1/\alpha_1$. Multiply all the α_i's (the first row) by this constant, and add to the bottom row. Our lineup now looks like this:

$$\begin{array}{cccc} \alpha_1 & \alpha_2 & \alpha_3 & \alpha_4 \\ \beta_1 & \beta_2 & \beta_3 & \beta_4 \\ 0 & \gamma_2 - \dfrac{\gamma_1 \alpha_2}{\alpha_1} & \gamma_3 - \dfrac{\gamma_1 \alpha_3}{\alpha_1} & \gamma_4 - \dfrac{\gamma_1 \alpha_4}{\alpha_1} \end{array}$$

In the same way, we can "annihilate" β_1 by finding a multiple of α_1 that when added to β_1 gives 0. By continuing this process (next annihilating γ_2), we tend toward a lineup that looks like this:

$$\begin{array}{cccc} \times & \times & \times & \times \\ 0 & \times & \times & \times \\ 0 & 0 & \times & \times \end{array}$$

A lineup like this, with no row all zeros when the lower left corner is all zeros, indicates that no linear combination of the α, β, γ vectors adds to zero, and the vectors are linearly independent. If, however, the lineup were to come out

$$\begin{array}{cccc} \times & \times & \times & \times \\ 0 & \times & \times & \times \\ 0 & 0 & 0 & 0 \end{array}$$

then there was some linear combination of the α, β, and γ vectors which added to zero (final last row), and the vectors would be linearly dependent.

This idea can now be expressed in formal terms.

Linear Algebra

> **Definition** A rectangular array of scalars (real or complex) is called a *matrix*. The scalar located in the ith row and jth column is called the i, j element of the matrix. A matrix with n rows and m columns is designated an $n \times m$ matrix.
>
> **Definition** The *elementary row operations* which may be performed on a matrix are
> 1. Multiplying a row by a constant.
> 2. Adding two rows together.
> 3. Exchanging two rows.
>
> **Definition** The *diagonal* (or, *principal diagonal*) of a matrix A consists of the elements $\{a_{ii}\}$, that is, $a_{11}, a_{22}, a_{33}, \cdots$, and so forth. These elements are called the *diagonal elements*.
>
> **Definition** A matrix, all of whose elements below and to the left of the diagonal are zero, is called a *triangular matrix*.
>
> **Definition** Matrices which may be formed from one another by elementary row operations are said to be *row equivalent*.
>
> **Theorem** A set of vectors is said to be linearly independent if the components of these vectors may be made the rows of a matrix which is row equivalent to a triangular matrix containing no all-zero rows.

This theorem expresses the method that we have outlined for testing for linear independence. The theorem and the method will be clearer with two examples.

EXAMPLE 1

Are the vectors $(1, -1, 3)$, $(2, -4, 1)$, $(0, 3, 2)$ linearly independent? We form the matrix (contained in parentheses) and perform elementary row operations on this matrix.

$$\begin{pmatrix} 1 & -1 & 3 \\ 2 & -4 & 1 \\ 0 & 3 & 2 \end{pmatrix}$$

The 3, 1 element is already zero. Force the 2, 1 element to be zero by adding to the second row -2 times the first row.

$$\xrightarrow[-2 \times \text{[1st row]}]{\text{[2nd row]}} \begin{pmatrix} 1 & -1 & 3 \\ 0 & -2 & -5 \\ 0 & 3 & 2 \end{pmatrix}$$

For simplicity, multiply the second row by $-\frac{1}{2}$.

$$\text{[2nd row]} \times (-\tfrac{1}{2}) \to \begin{pmatrix} 1 & -1 & 3 \\ 0 & 1 & \tfrac{5}{2} \\ 0 & 3 & 2 \end{pmatrix}$$

Force the 3, 2 element to be zero by multiplying the second row by -3 and adding to the third row.

$$\xrightarrow[-3 \times \text{[2nd row]}]{\text{[3rd row]}} \begin{pmatrix} 1 & -1 & 3 \\ 0 & 1 & \tfrac{5}{2} \\ 0 & 0 & -\tfrac{11}{2} \end{pmatrix}$$

The matrix is now in triangular form. All rows contain at least one non-zero element; therefore, the vectors are linearly independent.

EXAMPLE 2

Are the vectors $(1, 2i, 1 + i)$, $(4, 6 - i, 7)$, and $(-2, -6 + 5i, -5 + 2i)$ linearly independent? We again form the matrix and systematically perform elementary row operations.

$$\begin{pmatrix} 1 & 2i & 1+i \\ 4 & 6-i & 7 \\ -2 & -6+5i & -5+2i \end{pmatrix}$$

$$\text{[3rd]} + 2 \times \text{[1st]} \to \begin{pmatrix} 1 & 2i & 1+i \\ 4 & 6-i & 7 \\ 0 & -6+9i & -3+4i \end{pmatrix}$$

$$\text{[2nd]} - 4 \times \text{[1st]} \to \begin{pmatrix} 1 & 2i & 1+i \\ 0 & 6-9i & 3-4i \\ 0 & -6+9i & -3+4i \end{pmatrix}$$

$$\text{[3rd]} + \text{[2nd]} \to \begin{pmatrix} 1 & 2i & 1+i \\ 0 & 6-9i & 3-4i \\ 0 & 0 & 0 \end{pmatrix}$$

The matrix is now in triangular form. One row is all zeros; therefore, the vectors are linearly dependent. We can go back and find the constants that satisfy Eq. 3-8:

$$2\alpha^1 + \alpha^3 + (-4\alpha^1 + \alpha^2) = 0$$
$$-2\alpha^1 + \alpha^2 + \alpha^3 = 0$$

In this section we have developed the properties of vectors in a way parallel to those developed for functions. The Schmidt orthogonalization procedure, expansion in an orthonormal basis set, and the expansion theorem are results which may be borrowed from Chapter 2. A simple test for linear independence has been demon-

Linear Algebra

strated. This test introduces algebraic elements called matrices, and some of their properties and parts, such as elements, diagonals, rows and columns, and row equivalence.

3-2 MATRICES, DETERMINANTS, AND LINEAR EQUATIONS

Matrices. The definition of a matrix was introduced in the previous section as a formal aid to the test for linear independence of vectors. However, matrices have more applications than to serve as aids in this one kind of problem. From the point of view of quantum mechanics, matrices will be used to represent operators, as was hinted at in Chapter 1. This use of matrices will be studied in detail in Section 4 of this chapter, and prompts our consideration of more of the properties of matrices at this time.

> **Definition** Two matrices may be added if they have the same dimensions. The *sum* of an $n \times m$ matrix A and an $n \times m$ matrix B is an $n \times m$ matrix C whose elements are given by the formula
>
> $$c_{ij} = a_{ij} + b_{ij} \qquad (3\text{-}14)$$

That is, matrices are added by adding their elements. We shall notate matrices by capital letters, and their elements by lower case letters.

EXAMPLE

$$\begin{pmatrix} 1 & 3 & 2 & 5 \\ 0 & 7 & 9 & 4 \\ 6 & -2 & 5 & 1 \end{pmatrix} + \begin{pmatrix} 4 & -3 & 1 & -6 \\ 0 & 0 & -2 & 1 \\ 1 & -1 & 1 & -1 \end{pmatrix} = \begin{pmatrix} 5 & 0 & 3 & -1 \\ 0 & 7 & 7 & 5 \\ 7 & -3 & 6 & 0 \end{pmatrix}$$

The definition of addition is not one which surprises students. The definition of multiplication, however, does not resemble the usual idea of multiplying two numbers.

> **Definition** Two matrices may be multiplied if the multiplier has the same number of columns as the multiplicand has rows. The *product* of an $n \times m$ matrix A and an $m \times p$ matrix B is an $n \times p$ matrix C whose elements are given by the formula
>
> $$c_{ij} = \sum_{k=1}^{m} a_{ik} b_{kj} \qquad (3\text{-}15)$$

Notice the rule which restricts the matrices which can be multiplied.

$$\Big(n \times m\Big) \cdot \Big(m \times p\Big) = \Big(n \times p\Big)$$

In particular, square matrices of the same dimension may always be multiplied.

$$\Big(n \times n\Big) \cdot \Big(n \times n\Big) = \Big(n \times n\Big)$$

Since we shall be representing operators by square matrices, their properties are of particular importance to us.

Compare the way in which a product of two matrices is formed to the way in which a sum is formed. Matrices are added by adding their elements, but matrices are not multiplied simply by multiplying their elements. Although we shall have many occasions to use Eq. 3–15 per se, nevertheless, it is useful to see what that equation means in terms of the actual operations. Suppose we wish to compute the product of a 2×3 and a 3×4 matrix. The result is a 2×4 matrix. The 2, 3 element would be given by

$$c_{23} = \sum_{k=1}^{3} a_{2k} b_{k3} = a_{21} b_{13} + a_{22} b_{23} + a_{23} b_{33} \qquad (3\text{–}16)$$

A diagram might help visualize what elements these are.

Often people (even professionals) will multiply matrices by running their left forefinger across the row of the multiplier (left) matrix,

Linear Algebra

and their right forefinger down the column of the multiplicand (right) matrix.

EXAMPLE 1

As an example, find the product shown below·

$$\begin{pmatrix} 3 & 1 & 2 \\ 1 & 2 & 3 \end{pmatrix} \begin{pmatrix} 1 & 0 & 1 & 0 \\ 0 & 1 & 0 & 1 \\ 1 & 0 & 1 & 0 \end{pmatrix} = \begin{pmatrix} c_{11} & c_{12} & c_{13} & c_{14} \\ c_{21} & c_{22} & c_{23} & c_{24} \end{pmatrix}$$

By Eq. 3-15 we compute the eight matrix elements:

$$c_{11} = 3 \cdot 1 + 1 \cdot 0 + 2 \cdot 1 = 5 \qquad c_{21} = 1 \cdot 1 + 2 \cdot 0 + 3 \cdot 1 = 4$$
$$c_{12} = 3 \cdot 0 + 1 \cdot 1 + 2 \cdot 0 = 1 \qquad c_{22} = 1 \cdot 0 + 2 \cdot 1 + 3 \cdot 0 = 2$$
$$c_{13} = 3 \cdot 1 + 1 \cdot 0 + 2 \cdot 1 = 5 \qquad c_{23} = 1 \cdot 1 + 2 \cdot 0 + 3 \cdot 1 = 4$$
$$c_{14} = 3 \cdot 0 + 1 \cdot 1 + 2 \cdot 0 = 1 \qquad c_{24} = 1 \cdot 0 + 2 \cdot 1 + 3 \cdot 0 = 2$$

The product is

$$\begin{pmatrix} 5 & 1 & 5 & 1 \\ 4 & 2 & 4 & 2 \end{pmatrix}$$

The student should verify these elements, and also confirm the "forefinger method" for multiplying the two matrices shown.

EXAMPLE 2

Work out the product of two square matrices shown below.

$$\begin{pmatrix} 1 & 2 \\ 3 & 4 \end{pmatrix} \begin{pmatrix} 4 & 3 \\ 2 & 1 \end{pmatrix} = \begin{pmatrix} 8 & 5 \\ 20 & 13 \end{pmatrix}$$

Another peculiar property of matrix multiplication is that it is noncommutative. The multiplication of scalars, and the addition of scalars and matrices are all commutative operations: $ab = ba$, $a + b = b + a$, $A + B = B + A$. Matrix multiplication is, in general, not commutative: $AB \neq BA$. We can be concerned with commutation only for square matrices, since nonsquare matrices cannot be multiplied in both directions. In Example 2 above,

$$\begin{pmatrix} 1 & 2 \\ 3 & 4 \end{pmatrix} \begin{pmatrix} 4 & 3 \\ 2 & 1 \end{pmatrix} = \begin{pmatrix} 8 & 5 \\ 20 & 13 \end{pmatrix} \neq \begin{pmatrix} 13 & 20 \\ 5 & 8 \end{pmatrix} = \begin{pmatrix} 4 & 3 \\ 2 & 1 \end{pmatrix} \begin{pmatrix} 1 & 2 \\ 3 & 4 \end{pmatrix}$$

We continue our discussion of the elementary arithmetic of matrices by mentioning two matrices which are the analogs of the numbers zero and one. The *zero*, or *null matrix*, *0*, is a matrix, all of whose elements are zero. Analogous to the number zero, the zero matrix is the matrix that, when added to any matrix (of the same dimensions), gives the same matrix for the sum:

$$A + 0 = A \tag{3-17}$$

The *identity, or unit matrix*, E (denoted by some I or 1), is the (square) matrix whose diagonal elements are all ones, and whose off-diagonal elements are all zeros. This definition may be made very concise by the use of the Kronecker delta. We define

$$e_{ii} = 1 \tag{3-18a}$$

for all i and

$$e_{ij} = 0 \tag{3-18b}$$

for all $i \neq j$. This is equivalent to defining

$$e_{ij} = \delta_{ij} \tag{3-18c}$$

The identity matrix, defined only for square matrices, has a property analogous to the number one, namely, the product of E with any matrix (square and of the same dimension as E) gives that matrix as product: $EA = A$.

For practice in using the general formula, Eq. 3-15, we prove that E commutes with all matrices (of the same dimension). The i, j element of the product EA is

$$(EA)_{ij} = \sum_k e_{ik} a_{kj} = \sum_k \delta_{ik} a_{kj} = a_{ij} \tag{3-19}$$

but

$$(AE)_{ij} = \sum_k a_{ik} e_{kj} = \sum_k a_{ik} \delta_{kj} = a_{ij} = (EA)_{ij} \tag{3-20}$$

That there is a matrix with properties analogous to the number one prompts us to ask whether there is an analog to the reciprocal (and therefore to division) for matrices.

Definition The *inverse* of a square matrix A, denoted A^{-1}, is that matrix for which $A^{-1}A = AA^{-1} = E$. *Inverse* is defined only for square matrices.

We shall discuss two ways of finding the inverse of a matrix. The first of these is quite similar to the test for linear independence intro-

Linear Algebra 61

duced in the previous section. The second is defined in terms of determinants, and will be discussed somewhat later. The idea of row equivalence is the clue to the first method.

> **Theorem** A square matrix A has an inverse if and only if its rows are linearly independent. The inverse may be found by the following procedure. Perform elementary row operations on A to form E. Then, perform these same row operations in the same order on E to form A^{-1}.

Notice that a matrix, even a nonzero matrix, need not have an inverse.

As the theorem in the previous section implied, A is row equivalent to E only if the rows of A are linearly independent. The proof of this theorem is left to the interested reader to find in one of the many excellent texts on algebra. We shall illustrate the theorem by an example.

EXAMPLE
To find the inverse of

$$A = \begin{pmatrix} 1 & 2 \\ 3 & 4 \end{pmatrix}$$

we begin by performing elementary row operations on A to form E:

$$\begin{pmatrix} 1 & 2 \\ 3 & 4 \end{pmatrix} \xrightarrow{[2\text{nd}] - 3 \times [1\text{st}]} \begin{pmatrix} 1 & 2 \\ 0 & -2 \end{pmatrix} \xrightarrow{[2\text{nd}] \div -2} \begin{pmatrix} 1 & 2 \\ 0 & 1 \end{pmatrix} \xrightarrow{[1\text{st}] - 2 \times [2\text{nd}]} \begin{pmatrix} 1 & 0 \\ 0 & 1 \end{pmatrix} = E$$

We then perform these same operations on E:

$$\begin{pmatrix} 1 & 0 \\ 0 & 1 \end{pmatrix} \xrightarrow{[2\text{nd}] - 3 \times [1\text{st}]} \begin{pmatrix} 1 & 0 \\ -3 & 1 \end{pmatrix} \xrightarrow{[2\text{nd}] \div -2}$$

$$\begin{pmatrix} 1 & 0 \\ \frac{3}{2} & -\frac{1}{2} \end{pmatrix} \xrightarrow{[1\text{st}] - 2 \times [2\text{nd}]} \begin{pmatrix} -2 & 1 \\ \frac{3}{2} & -\frac{1}{2} \end{pmatrix} = A^{-1}$$

As a check, multiply the answer by A:

$$\begin{pmatrix} -2 & 1 \\ \frac{3}{2} & -\frac{1}{2} \end{pmatrix} \begin{pmatrix} 1 & 2 \\ 3 & 4 \end{pmatrix} = \begin{pmatrix} 1 & 0 \\ 0 & 1 \end{pmatrix}$$

This technique is cumbersome and prone to error; the student will find the technique using determinants to be somewhat safer. We conclude our discussion of matrix algebra with one further definition.

Definition The *transpose* of a square matrix A, denoted A', has elements $a'_{ij} = a_{ji}$. It can be visualized by flipping the matrix A about its diagonal.

The transpose of a matrix plays an important role in quantum mechanics. Multiplication of transposes follows a simple rule,

$$(AB)' = B'A' \tag{3-21}$$

which is also followed by inverses when they exist,

$$(AB)^{-1} = B^{-1}A^{-1} \tag{3-22}$$

Determinants. The formal definition of a determinant is sufficiently complicated that we begin with a practical example of the use of determinants. One of the important applications of determinants is in the solution of simultaneous linear equations. As an example, consider this set of equations:

$$\begin{aligned} a_{11}x_1 + a_{12}x_2 &= b_1 \\ a_{21}x_1 + a_{22}x_2 &= b_2 \end{aligned} \tag{3-23}$$

Writing the equations in this form does, of course, prejudice us toward a matrix point of view which we shall develop later. For the present, we solve Eqs. 3-23 by substitution. We find from the second equation that $x_2 = (b_2 - a_{21}x_1)/a_{22}$, and, substituting this result into the first equation, we obtain the solution

$$x_1 = \frac{a_{22}b_1 - a_{12}b_2}{a_{11}a_{22} - a_{12}a_{21}} \tag{3-24}$$

The denominator of this expression contains all four of the coefficients in the simultaneous equations 3-23. In fact, this denominator is the *determinant* of a matrix which we shall call the *coefficient matrix*.

$$A = \begin{pmatrix} a_{11} & a_{12} \\ a_{21} & a_{22} \end{pmatrix} \tag{3-25a}$$

$$\begin{vmatrix} a_{11} & a_{12} \\ a_{21} & a_{22} \end{vmatrix} \equiv a_{11}a_{22} - a_{12}a_{21} \tag{3-25b}$$

The numerator of Eq. 3-24 may also be written as a determinant, so that the results are

Linear Algebra

$$x_1 = \frac{\begin{vmatrix} b_1 & a_{12} \\ b_2 & a_{22} \end{vmatrix}}{\begin{vmatrix} a_{11} & a_{12} \\ a_{21} & a_{22} \end{vmatrix}} \qquad x_2 = \frac{\begin{vmatrix} a_{11} & b_1 \\ a_{21} & b_2 \end{vmatrix}}{\begin{vmatrix} a_{11} & a_{12} \\ a_{21} & a_{22} \end{vmatrix}} \qquad (3\text{-}26)$$

This result is a special case of a theorem that we shall discuss later called *Cramer's rule*. The 2 × 2 determinant has the simple definition given in Eq. 3-25b. Determinants corresponding to large square matrices have a definition that is more complex, and not immediately generalizable from Eq. 3-25b.

> **Definition** A *permutation of n integers*, P, is a way of ordering the n integers; there are $n!$ such permutations. The *sign of a permutation*, sign P, is positive if an even number of interchanges of two indices achieves the permutation, and negative if an odd number of interchanges achieves the permutation.

EXAMPLE

For the integers (1, 2, 3), an odd permutation would be (2, 1, 3); an even permutation would be (2, 3, 1).

> **Definition** The *determinant* of a square matrix A, denoted $|A|$ or det A, is a number, real or complex, such that
>
> $$|A| = \sum_{\text{all perm. } P}^{n!} (\text{sign } P) a_{1,P_1} a_{2,P_2} \cdots a_{n,P_n} \qquad (3\text{-}27)$$

EXAMPLE

As an example of this formula for finding determinants, consider a 3 × 3 example. There are 3! = 6 terms. Evaluate

$$\begin{vmatrix} a_{11} & a_{12} & a_{13} \\ a_{21} & a_{22} & a_{23} \\ a_{31} & a_{32} & a_{33} \end{vmatrix}$$

Term	Permutation	Interchanges	Sign P
$a_{11}a_{22}a_{33}$	1 2 3	0	+
$a_{11}a_{23}a_{32}$	1 3 2	1	−
$a_{12}a_{21}a_{33}$	2 1 3	1	−
$a_{12}a_{23}a_{31}$	2 3 1	2	+
$a_{13}a_{22}a_{31}$	3 2 1	1	−
$a_{13}a_{21}a_{32}$	3 1 2	2	+

Putting these terms together, with the correct signs, we get

$$|A| = a_{11}a_{22}a_{33} + a_{12}a_{23}a_{31} + a_{13}a_{21}a_{32} - a_{11}a_{23}a_{32}$$
$$- a_{12}a_{21}a_{33} - a_{13}a_{22}a_{31} \qquad (3\text{-}28)$$

This is a cumbersome procedure at best, and there is a simpler way. We may get a clue to this simpler way by grouping together all the terms which contain a_{11}, a_{12}, and a_{13}, respectively.

$$|A| = a_{11}(a_{22}a_{33} - a_{23}a_{32}) + a_{12}(a_{23}a_{31} - a_{21}a_{33})$$
$$+ a_{13}(a_{21}a_{32} - a_{22}a_{31}) \qquad (3\text{-}29a)$$

$$= a_{11}\begin{vmatrix} a_{22} & a_{23} \\ a_{32} & a_{33} \end{vmatrix} - a_{12}\begin{vmatrix} a_{21} & a_{23} \\ a_{31} & a_{33} \end{vmatrix} + a_{13}\begin{vmatrix} a_{21} & a_{22} \\ a_{31} & a_{32} \end{vmatrix} \qquad (3\text{-}29b)$$

Eq. 3–29a illustrates the *expansion of a determinant in cofactors;* Eq. 3–29b, the *expansion of a determinant in minors.*

> **Definition** The *cofactor*, A_{ij}, of an element a_{ij} in an $n \times n$ square matrix A is a number that can be represented by an $(n-1) \times (n-1)$ determinant, such that the determinant of A is given by
>
> $$|A| = \sum_j a_{ij} A_{ij}$$
>
> for any and all i. The *minor*, M_{ij}, of an element a_{ij} in an $n \times n$ square matrix A is a number, represented by an $(n-1) \times (n-1)$ determinant found by striking the ith row and jth column from A. The minor and the cofactor are related by at most a change of sign: $A_{ij} = M_{ij}(-1)^{i+j}$.

EXAMPLE

As an example, consider the 2, 3 minor and cofactor for this 4×4 matrix:

$$A = \begin{pmatrix} a_{11} & a_{12} & a_{13} & a_{14} \\ a_{21} & a_{22} & a_{23} & a_{24} \\ a_{31} & a_{32} & a_{33} & a_{34} \\ a_{41} & a_{42} & a_{43} & a_{44} \end{pmatrix} \text{---strike out 2nd row------}$$

strike out 3rd column

$$M_{23} = \begin{vmatrix} a_{11} & a_{12} & a_{14} \\ a_{31} & a_{32} & a_{34} \\ a_{41} & a_{42} & a_{44} \end{vmatrix}$$

$$A_{23} = -\begin{vmatrix} a_{11} & a_{12} & a_{14} \\ a_{31} & a_{32} & a_{34} \\ a_{41} & a_{42} & a_{44} \end{vmatrix} = (-1)^{2+3} M_{23}$$

Linear Algebra

Theorem The determinant of an $n \times n$ matrix A may be expanded in cofactors or minors:

$$\det A = \sum_j a_{ij}A_{ij} = \sum_j a_{ij}(-1)^{i+j}M_{ij} \qquad (3\text{-}30)$$

for any i.

EXAMPLE

Evaluate the 3×3 determinant shown below by expansion in minors and by direct use of Eq. 3-28.

$$\begin{vmatrix} 1 & 0 & 2 \\ 3 & 4 & -1 \\ -2 & 1 & 1 \end{vmatrix} = 1 \begin{vmatrix} 4 & -1 \\ 1 & 1 \end{vmatrix} - 0 \begin{vmatrix} 3 & -1 \\ -2 & 1 \end{vmatrix} + 2 \begin{vmatrix} 3 & 4 \\ -2 & 1 \end{vmatrix} = 5 - 0 + 22 = 27$$

$$= -3 \begin{vmatrix} 0 & 2 \\ 1 & 1 \end{vmatrix} + 4 \begin{vmatrix} 1 & 2 \\ -2 & 1 \end{vmatrix} - (-1) \begin{vmatrix} 1 & 0 \\ -2 & 1 \end{vmatrix} = 6 + 20 + 1 = 27$$

$$= -2 \begin{vmatrix} 0 & 2 \\ 4 & -1 \end{vmatrix} - 1 \begin{vmatrix} 1 & 2 \\ 3 & -1 \end{vmatrix} + 1 \begin{vmatrix} 1 & 0 \\ 3 & 4 \end{vmatrix} = 16 + 7 + 4 = 27$$

$$= (1)(4)(1) + (0)(-1)(-2) + (2)(3)(1) - (1)(-1)(1) - (0)(3)(1) - (2)(4)(-2) = 4 + 0 + 6 + 1 - 0 + 16 = 27$$

The first equation in the example is an expansion in minors of the first row; the second, of the second row; the third, of the third row. The last equation is a direct evaluation, using Eq. 3-28, the special case of Eq. 3-27.

Expansion by minors is especially helpful if there are zeros in the determinant, as in the first equation of the previous example.

The elementary row operations that we have used to test for linear independence in vectors may also be used to aid in the evaluation of determinants.

Theorem Multiplying a row of a square matrix by a constant multiplies the determinant of that matrix by that constant.

If we expand A in cofactors of the row in question,

$$|A| = \sum_j a_{ij}A_{ij} \qquad (3\text{-}31)$$

Multiplying the ith row by a constant c gives a new matrix B, whose determinant is

$$|B| = \sum_j b_{ij}B_{ij} = \sum_j ca_{ij}A_{ij} = c\sum_j a_{ij}A_{ij} = c|A| \quad (3\text{-}32)$$

Theorem The exchange of two rows of a matrix changes the sign of the determinant of that matrix.

This result can be proved by relying on the general definition of a determinant. Without going into details, all the terms in the expansion are the same for the original matrix and the matrix with the exchanged rows. The signs of the terms will be different throughout because one additional exchange of indices has been made. This theorem has an immediate corollary.

Corollary A square matrix with two identical rows has a zero determinant.

This is so because exchanging the identical rows gives a determinant of opposite sign, but exchanging the identical rows must also give the same determinant. Hence $|A| = -|A|$, which is true only if $|A| = 0$. Finally, the result of performing the last elementary row operation is obtained.

Theorem The addition of two rows of a square matrix leaves the determinant unchanged.

We may prove this by expanding the determinant of B, where the sum of the ith and kth rows of A forms the ith row of B. Use cofactors of the ith row of B.

$$|B| = \sum_j b_{ij}B_{ij} = \sum_j (a_{ij} + a_{kj})A_{ij} = \sum_j a_{ij}A_{ij} + \sum_j a_{kj}A_{ij}$$

$$(3\text{-}33)$$

The first term is the expansion of $|A|$ in cofactors of the ith row of A. The second term looks like an expansion in cofactors—it would be, if the matrix A had identical ith and kth rows. Hence the second term represents a cofactor expansion of a determinant with two identical rows; that determinant is zero. Therefore, $|B| = |A|$.

By way of recapitulation, the accompanying table gives the results of the elementary row operations on the value of a determinant.

Linear Algebra

Elementary row operation	Effect on value of determinant
1. Multiply row by constant.	Multiply determinant by constant.
2. Exchange rows.	Change sign or determinant.
3. Add rows.	No change.

We conclude this discussion of elementary properties of determinants with two further results of importance.

> **Theorem** The determinant of the product of two square matrices is the product of their determinants: $|AB| = |A| \cdot |B|$.
>
> **Theorem** The determinant of a square matrix is equal to the determinant of the transpose of that matrix: $|A| = |A'|$.

The proof of the latter theorem relies on an examination of the general definition of a determinant. All the terms are alike, and all the signs match as well. This theorem allows us to rephrase all our results about matrices in a column language rather than a row language. These statements are given in the following list, which also serves to sum up our knowledge of determinants.

Row-language statements	Analogous column-language statements
1. Expansion of a determinant by row cofactors or minors: $$|A| = \sum_j a_{ij}A_{ij} = \sum_j a_{ij}M_{ij}(-1)^{i+j}.$$	Expansion of a determinant by column cofactors or minors: $$|A| = \sum_i a_{ij}A_{ij} = \sum_i a_{ij}M_{ij}(-1)^{i+j}.$$
2. Multiply row by constant, multiply determinant by constant.	Multiply column by constant, multiply determinant by constant.
3. Interchange rows, change sign of determinant.	Interchange columns, change sign of determinant.
4. Matrix with two identical rows has zero determinant.	Matrix with two identical columns has zero determinant.
5. Adding two rows together leaves determinant unchanged.	Adding two columns together leaves determinant unchanged.

With this background in the properties of determinants, we may approach two problems in a simpler way. Both the problem of testing for linear independence of vectors (as rows of a matrix) and the problem of matrix inversion have been solved using row equivalence. Each of these procedures is cumbersome, especially for large matrices. A straightforward application of the properties of determinants affords a more compact solution to both of these problems.

Theorem A square matrix A has an inverse if $|A| \neq 0$; the i, j element of the inverse matrix A^{-1} is

$$(A^{-1})_{ij} = \frac{A'_{ij}}{|A|} = \frac{A_{ji}}{|A|} \tag{3-34}$$

Expanding $|A|$ in cofactors, we have

$$a_{i1}A_{i1} + a_{i2}A_{i2} + \cdots + a_{in}A_{in} = |A| = \sum_j a_{ij}A_{ij} \tag{3-35}$$

but, using the same principle as we did in Eq. 3-33 to set the second term equal to zero, we have also

$$\sum_j a_{kj}A_{ij} = 0 \tag{3-36}$$

for $i \neq k$. These may be combined as

$$\sum_j a_{kj} \frac{A_{ij}}{|A|} = \delta_{ki} \tag{3-37}$$

where δ_{ik} is an element of the unit matrix E. The left side of Eq. 3-37 looks almost like a matrix product. We may make it be such by writing, for the cofactor A_{ij}, the cofactor A'_{ji} from the transposed matrix. Then the element $A'_{ji}/|A| = a_{ji}^{-1}$, since

$$\sum_j a_{kj} a_{ji}^{-1} = \delta_{ki}.$$

Notice that these elements exist only for nonzero $|A|$. Here, the symbol a_{ij}^{-1} is not used to represent $1/a_{ij}$, but to represent the i, j element of the matrix A^{-1}. This affords a more straightforward path to matrix inversion. As an example, let us do the previous example again.

Linear Algebra

EXAMPLE
Find the inverse of
$$\begin{pmatrix} 1 & 2 \\ 3 & 4 \end{pmatrix}$$

By the formula given above,

$$a_{11}^{-1} = \frac{A'_{11}}{|A|} = \frac{4}{-2} = -2$$

$$a_{12}^{-1} = \frac{A'_{12}}{|A|} = \frac{-2}{-2} = 1$$

$$a_{21}^{-1} = \frac{A'_{21}}{|A|} = \frac{-3}{-2} = \frac{3}{2}$$

$$a_{22}^{-1} = \frac{A'_{22}}{|A|} = \frac{1}{-2} = \frac{-1}{2}$$

hence

$$A^{-1} = \begin{pmatrix} -2 & 1 \\ \frac{3}{2} & -\frac{1}{2} \end{pmatrix}$$

as was found previously.

Finally, we may combine our result on linear independence and matrix inversion (derived by row-equivalence arguments) with our result on nonzero determinants and matrix inversion to yield the following result relating linear independence to nonzero determinants.

Theorem The rows of a square matrix are linearly independent if and only if the determinant of the matrix is nonzero.

Simultaneous Linear Equations. The methods of matrix and determinant algebra may be applied with force to the solution of simultaneous linear equations. The results will be stated without proof, and a number of examples will illustrate the methods.

By way of introduction to this problem, note that the general set of simultaneous linear equations may be represented as a single matrix equation. A set of m equations in n unknowns may be represented either by

$$\begin{array}{c} a_{11}x_1 + a_{12}x_2 + \cdots + a_{1n}x_n = b_1 \\ a_{21}x_1 + a_{22}x_2 + \cdots + a_{2n}x_n = b_2 \\ \vdots \qquad \vdots \qquad \qquad \vdots \qquad \vdots \\ a_{m1}x_1 + a_{m2}x_2 + \cdots + a_{mn}x_n = b_m \end{array} \qquad (3\text{-}38)$$

or by the matrix equation

$$AX = B \qquad (3\text{-}39)$$

where A is an $m \times n$ matrix of coefficients, X an $n \times 1$ matrix of unknowns, and B an $m \times 1$ matrix of constants:

$$\begin{pmatrix} a_{11} & a_{12} & \cdots & a_{1n} \\ a_{21} & a_{22} & \cdots & a_{2n} \\ \vdots & \vdots & & \vdots \\ a_{m1} & a_{m2} & \cdots & a_{mn} \end{pmatrix} \begin{pmatrix} x_1 \\ x_2 \\ \vdots \\ x_n \end{pmatrix} = \begin{pmatrix} b_1 \\ b_2 \\ \vdots \\ b_m \end{pmatrix} \qquad (3\text{-}40)$$

We shall define at the outset two concepts.

Definition The *augmented matrix* for a set of m simultaneous linear equations in n unknowns is an $m \times (n + 1)$ matrix formed by appending the $m \times 1$ matrix of constants on the right of the $n \times m$ matrix of coefficients. If, for example, A and B are the matrices in Eq. 3-40, then the augmented matrix $*A$ is

$$*A = \begin{pmatrix} a_{11} & a_{12} & \cdots & a_{1n} & b_1 \\ a_{21} & a_{22} & \cdots & a_{2n} & b_2 \\ \vdots & \vdots & & \vdots & \vdots \\ a_{m1} & a_{m2} & \cdots & a_{mn} & b_m \end{pmatrix} \qquad (3\text{-}41)$$

Definition A set of simultaneous linear equations is called *homogeneous* if the matrix B is zero.

The results, offered almost entirely without proof, are expressed in the following theorems.

Theorem A set of n simultaneous linear equations in n unknowns has a solution if the coefficient determinant is not zero. The solution may be constructed by matrix inversion, or by *Cramer's rule*.

If $AX = B$, then $A^{-1}AX = A^{-1}B$. So the entire set of solutions X may be generated if A^{-1} is known. However, A^{-1} exists only if $|A| \neq 0$. The result is

$$X = A^{-1}B \qquad (3\text{-}42)$$

In particular,

$$x_i = \sum_j (A^{-1})_{ij} b_j$$

Linear Algebra

$$= \sum_j \frac{A_{ji}}{|A|} b_j = \sum_j \frac{b_j A_{ji}}{|A|} \quad (3\text{-}43)$$

Equation 3–43 involves a sum,

$$\sum_j b_j A_{ji}$$

which is the column cofactor expansion of a determinant whose ith column is the constant matrix B. However, this is Cramer's rule in its familiar form:

$$x_i = \frac{\begin{vmatrix} a_{11} & \cdots & b_1 & \cdots & a_{1n} \\ a_{21} & \cdots & b_2 & \cdots & a_{2n} \\ \vdots & & \vdots & & \vdots \\ a_{n1} & \cdots & b_n & \cdots & a_{nn} \end{vmatrix}}{|A|} \quad (3\text{-}44)$$

Theorem A set of m simultaneous linear equations in n unknowns, $AX = B$, has a solution if the number of linearly independent rows, r, in A is the same as the number of linearly independent rows in $*A$. If such is the case, r of the unknowns may be expressed in terms of the remaining $n - r$, which may be given arbitrary values.

Theorem A set of m simultaneous, linear, homogeneous equations in n unknowns always has the trivial solution $X = 0$. There will be a nonzero (nontrivial) solution if the coefficient matrix A has fewer linearly independent rows than there are unknowns. Again, r of the unknowns may be expressed in terms of the remaining $n - r$, which may be assigned arbitrary values.

We conclude this section with a number of examples of these theorems.

EXAMPLE 1
Two linear equations in two unknowns.

$$4x + 4y = 2$$
$$8x - 2y = 4$$

The Cramer's rule solution is

$$x = \frac{\begin{vmatrix} 2 & 4 \\ 4 & -2 \end{vmatrix}}{\begin{vmatrix} 4 & 4 \\ 8 & -2 \end{vmatrix}} = \frac{-20}{-40} = \frac{1}{2}$$

$$y = \frac{\begin{vmatrix} 4 & 2 \\ 8 & -4 \end{vmatrix}}{\begin{vmatrix} 4 & 4 \\ 8 & -2 \end{vmatrix}} = \frac{0}{-40} = 0$$

The matrix solution can be found from the rule that, if $AX = C$, $X = A^{-1}C$. The inverse of the coefficient matrix is the matrix

$$\begin{pmatrix} \frac{1}{20} & \frac{1}{10} \\ \frac{1}{5} & -\frac{1}{10} \end{pmatrix}$$

Then the solution to the set of equations is

$$\begin{pmatrix} x \\ y \end{pmatrix} = \begin{pmatrix} \frac{1}{20} & \frac{1}{10} \\ \frac{1}{5} & -\frac{1}{10} \end{pmatrix} \begin{pmatrix} 2 \\ 4 \end{pmatrix} = \begin{pmatrix} \frac{2}{20} + \frac{4}{10} \\ \frac{2}{5} - \frac{4}{10} \end{pmatrix} = \begin{pmatrix} \frac{1}{2} \\ 0 \end{pmatrix}$$

This matrix equation says $x = \frac{1}{2}$, $y = 0$.

EXAMPLE 2

Three equations in two unknowns.

$$4x + 4y = 2$$
$$8x - 2y = 4$$
$$x + y = 1$$

The matrix of coefficients,

$$\begin{pmatrix} 4 & 4 \\ 8 & -2 \\ 1 & 1 \end{pmatrix}$$

is row equivalent to

$$\begin{pmatrix} 1 & 1 \\ 0 & 10 \\ 0 & 0 \end{pmatrix}$$

and has two linearly independent rows. The augmented matrix,

$$\begin{pmatrix} 4 & 4 & 2 \\ 8 & -2 & 4 \\ 1 & 1 & 1 \end{pmatrix}$$

Linear Algebra

is row equivalent to

$$\begin{pmatrix} 1 & 1 & \frac{1}{2} \\ 0 & 1 & 0 \\ 0 & 0 & 1 \end{pmatrix}$$

and has three linearly independent rows. There is, therefore, no solution to this set of equations.

EXAMPLE 3

Three equations in two unknowns.

$$4x + 4y = 2$$
$$8x - 2y = 4$$
$$3x + \frac{y}{2} = \frac{3}{2}$$

The matrix of coefficients,

$$\begin{pmatrix} 4 & 4 \\ 8 & -2 \\ 3 & \frac{1}{2} \end{pmatrix}$$

is row equivalent to

$$\begin{pmatrix} 1 & 1 \\ 0 & 1 \\ 0 & 0 \end{pmatrix}$$

and has two linearly independent rows. The augmented matrix

$$\begin{pmatrix} 4 & 4 & 2 \\ 8 & -2 & 4 \\ 3 & \frac{1}{2} & \frac{3}{2} \end{pmatrix}$$

is row equivalent to

$$\begin{pmatrix} 1 & 1 & \frac{1}{2} \\ 0 & 1 & 0 \\ 0 & 0 & 0 \end{pmatrix}$$

and also has two linearly independent rows. This set of equations does have a solution. The solution can be obtained from any matrix row equivalent to the augmented matrix. Using this last matrix as a guide, write the *equivalent* matrix equation

$$\begin{pmatrix} 1 & 1 \\ 0 & 1 \end{pmatrix} \begin{pmatrix} x \\ y \end{pmatrix} = \begin{pmatrix} \frac{1}{2} \\ 0 \end{pmatrix}$$

The solution to this equation can be found by the usual technique (see Example 1), and is $x = \frac{1}{2}$, $y = 0$, which is the same as the solution to Example 1. This could also be expressed by saying that, since the equations of this example are linearly dependent, and since two of these equations are identical to the equations of Example 1, the solution to this set must also be identical to the solution to Example 1.

EXAMPLE 4

Two equations in three unknowns.

$$x + y + z = 2$$
$$x - y - z = 1$$

The coefficient matrix,

$$\begin{pmatrix} 1 & 1 & 1 \\ 1 & -1 & -1 \end{pmatrix}$$

has two linearly independent rows, since it is row equivalent to the matrix

$$\begin{pmatrix} 1 & 1 & 1 \\ 0 & -2 & -2 \end{pmatrix}$$

The augmented matrix,

$$\begin{pmatrix} 1 & 1 & 1 & 2 \\ 1 & -1 & -1 & 1 \end{pmatrix}$$

which is row equivalent to

$$\begin{pmatrix} 1 & 1 & 1 & 2 \\ 0 & -2 & -2 & -1 \end{pmatrix}$$

also has two linearly independent rows; therefore, this set of equations has a solution. As in the previous example, this set of equations can be solved by appealing to a matrix which is row equivalent to the augmented matrix, and writing

$$\begin{pmatrix} 1 & 1 & 1 \\ 0 & -2 & -2 \end{pmatrix} \begin{pmatrix} x \\ y \\ z \end{pmatrix} = \begin{pmatrix} x + y + z \\ -2y - 2z \end{pmatrix} = \begin{pmatrix} 2 \\ -1 \end{pmatrix}$$

whence, $x = 2 - y - z$ and $y + z = \frac{1}{2}$. The solution can be expressed $x = \frac{3}{2}$, $z = \frac{1}{2} - y$. There are an infinite number of solutions. That is, there is

Linear Algebra

one solution for every possible value of y. The general solution can be checked by substitution into the equations themselves, which gives an identity in each case. One solution might be $x = \frac{3}{2}$, $y = 0$, $z = \frac{1}{2}$; another might be $x = \frac{3}{2}$, $y = 1$, $z = -\frac{1}{2}$, and so forth.

EXAMPLE 5

Two equations in three unknowns.

$$x + y + z = 2$$
$$x + y + z = 1$$

This set of equations has no solutions, which is obvious by inspection. The result can be proved by examining the coefficient matrix,

$$\begin{pmatrix} 1 & 1 & 1 \\ 1 & 1 & 1 \end{pmatrix}$$

which is row equivalent to

$$\begin{pmatrix} 1 & 1 & 1 \\ 0 & 0 & 0 \end{pmatrix}$$

and therefore has one linearly independent row. On the other hand, the augmented matrix,

$$\begin{pmatrix} 1 & 1 & 1 & 2 \\ 1 & 1 & 1 & 1 \end{pmatrix}$$

is row equivalent to

$$\begin{pmatrix} 1 & 1 & 1 & 2 \\ 0 & 0 & 0 & -1 \end{pmatrix}$$

and therefore has two linearly independent rows. Since the coefficient matrix and the augmented matrix have different numbers of linearly independent rows, there is no solution to this set of equations.

EXAMPLE 6

Three homogeneous equations in three unknowns.

$$3x + 4y + z = 0$$
$$2x + 6y + 4z = 0$$
$$x - y + z = 0$$

The coefficient matrix,

$$\begin{pmatrix} 3 & 4 & 1 \\ 2 & 6 & 4 \\ 1 & -1 & 1 \end{pmatrix}$$

has a determinant equal to 30, and three linearly independent rows, and therefore there is no solution to this set of homogeneous equations, except the trivial solution $x = y = z = 0$.

EXAMPLE 7

Three homogeneous equations in three unknowns.

$$x + y + z = 0$$
$$x - y - z = 0$$
$$x + 3y + 3z = 0$$

The coefficient matrix has a determinant

$$\begin{vmatrix} 1 & 1 & 1 \\ 1 & -1 & -1 \\ 1 & 3 & 3 \end{vmatrix} = 0$$

and there is a solution to this set of equations. To find the solution, rely again on a matrix which is row equivalent to the augmented matrix. The augmented matrix is

$$\begin{pmatrix} 1 & 1 & 1 & 0 \\ 1 & -1 & -1 & 0 \\ 1 & 3 & 3 & 0 \end{pmatrix}$$

which is row eqiuvalent to

$$\begin{pmatrix} 1 & 1 & 1 & 0 \\ 0 & 1 & 1 & 0 \\ 0 & 0 & 0 & 0 \end{pmatrix}$$

and has only two linearly independent rows. The solution of a matrix equation written from this augmented matrix is $x + y + z = 0$, and $y + z = 0$. That is to say, a general solution is $x = 0$, $y = -z$. There are any number of particular solutions, as before. One is the ever present trivial solution $x = y = z = 0$; another might be $x = 0, y = 1, z = -1$.

3-3 LINEAR TRANSFORMATIONS

In defining a function in Chapter 2, we emphasized the single value that the function delivered if presented with a single value of the independent variable on some specified interval. A transformation is a generalization of this concept.

Linear Algebra

Definition Let there exist n independent variables x_i ($i = 1, 2, \ldots, n$), each defined on a specified interval, such as $a_1 \leq x_1 \leq b_1$, $a_2 \leq x_2 \leq b_2$, and so forth. If there then exist m dependent variables, y_i, each of which is a single-valued function of the n independent variables x_i, we say that there exists a *transformation* which carries a region in n-dimensional space (or n space) into m-dimensional space (or m space). We write $y_i = T(x_i)$, and refer to the set of values y_i as the *image* of the set of values x_i under T. We shall represent transformations with capital letters.

To continue the algebraic theme established in the first sections of this chapter, we restrict the discussion at once to linear transformations.

Definition A transformation A is linear if
1. $A(x_i + x_j) = A(x_i) + A(x_j)$
2. $A(cx_i) = cA(x_i)$

where c is a scalar.

This definition contains two aspects of linearity which should be familiar in retrospect. These two aspects necessarily imply that a linear transformation has the form

$$\begin{aligned} a_{11}x_1 + a_{12}x_2 + \cdots + a_{1n}x_n &= y_1 \\ \vdots \qquad\qquad \vdots \qquad\qquad \vdots &\quad \vdots \\ a_{m1}x_1 + a_{m2}x_2 + \cdots + a_{mn}x_n &= y_m \end{aligned} \qquad (3\text{--}45)$$

which, expressed in matrix language, is

$$AX = Y \qquad (3\text{--}46)$$

where A is an $m \times n$ matrix, X an $n \times 1$ matrix, and Y an $m \times 1$ matrix. Having discovered that matrices may be used to represent linear transformations, we may call upon our past development of matrix algebra to help understand the properties of these transformations. The introduction of the term *rank* will simplify the discussion.

Definition The *rank of a matrix* is the largest number of linearly independent rows in the matrix; alternatively, the rank of a matrix is the dimension of the largest nonzero determinant which can be formed from the matrix elements. The *rank of a linear transformation* is equal to the rank of the matrix which represents the linear transformation.

The relationship between nonzero determinants and linear independence establishes the alternative definition of rank. We shall now present a general result and then illustrate it with a specific example.

Theorem Suppose a linear transformation of n space into m space has rank r. The image in m space will be an r-dimensional subregion which is described by a linear equation in the m coordinates of m space. In particular, the transformation will map n space into all of m space if the transformation has rank $r = m$.

By way of proof, consider the $m \times n$ matrix which represents a linear transformation. If only r of the rows of this matrix are linearly independent, then $m - r$ of the rows are dependent. This implies that $m - r$ of the coordinates in m space are linearly dependent. As a consequence, the image of n space in m space is on an r-dimensional subregion whose equation is linear in the m coordinates.

EXAMPLE
Consider the transformation

$$\left. \begin{array}{l} x + y + z = u \\ x + y + z = v \end{array} \right\} T$$

The rank of T is the rank of the matrix

$$\begin{pmatrix} 1 & 1 & 1 \\ 1 & 1 & 1 \end{pmatrix}$$

which has one linearly independent row; the rank is therefore one. In fact, since the rows of the matrix are the same, $u = v$. Hence the image of XYZ space on UV space is the line $u = v$, a one-dimensional subregion in UV space.

In quantum mechanics, where, as the student will learn, a premium is placed on the properties of orthogonality and normality of functions or vectors, particular attention is paid to the kinds of transformations which retain these properties. In a Euclidean vector space, this kind of transformation is called *orthogonal;* in a Hermitian vector space, the analogous transformation is called *unitary*.

Linear Algebra

> **Definition** An *orthogonal transformation* preserves the length (normalization) and orthogonality of vectors in a Euclidean vector space; a *unitary transformation* preserves orthogonality and normalization of vectors in a Hermitian vector space.

This definition is basic, but rather sterile unless we interpret it in terms of the structure of the transformation—the relations between the matrix elements of the matrix representing the transformation. We may examine this structure by using this very definition. Suppose there is a complete, orthonormal set of vectors $\{\phi^i\}$. We have already established that an arbitrary vector α may be expanded in terms of the vectors $\{\phi^i\}$ according to

$$\alpha = \sum_i a_i \phi^i \qquad (3\text{-}47)$$

as described in Eq. 3-11 and the discussion there. Now suppose we require that the expansion of α in terms of the ϕ^i's preserve length and orthogonality. That is, suppose α is a member of a second complete orthonormal set $\{\psi^i\}$, $\alpha = \psi^j$, say. Then

$$\psi^j = \sum_i a_{ji} \phi^i \qquad (3\text{-}48)$$

The coefficient a_i in Eq. 3-47 has now been modified with a second index j indicating which of the vectors in the set $\{\psi^j\}$ is being considered. The inverse expansion is also possible (this is to be proved in one of the problems), so that

$$\phi^i = \sum_k b_{ik} \psi^k \qquad (3\text{-}49)$$

Combining these equations, we find

$$\psi^j = \sum_i a_{ji} \phi^i = \sum_i a_{ji} \sum_k b_{ik} \psi^k = \sum_k \left(\sum_i a_{ji} b_{ik} \right) \psi^k \qquad (3\text{-}50)$$

The only way for Eq. 3-50 to be true, since the ψ^i's are linearly independent, is for $\sum_i a_{ji} b_{ik} = 1$ for $j = k$, and $\sum_i a_{ji} b_{ik} = 0$ for $j \neq k$. This then gives

$$\sum_i a_{ji}b_{ik} = \delta_{jk} \tag{3-51}$$

a statement that, if expressed not in the language of expansion coefficients but in the language of matrices, becomes

$$AB = E \tag{3-52}$$

or

$$A = B^{-1} \tag{3-53}$$

This should not be surprising. The transformation from the basis set $\{\phi^i\}$ to the basis set $\{\psi^i\}$ must be representable by a square matrix with nonzero determinant; hence, its inverse must exist. It is only reasonable that the inverse transformation be directly related to the reverse expansion of vectors.

We have not yet drawn upon the property that the orthonormality of the vectors be preserved under this transformation. Equation 3-53 is a result of requiring that completeness be preserved. For orthonormality of the ψ^i's, we have, in a Euclidean vector space,

$$\langle \psi^i \mid \psi^j \rangle = \delta_{ij} = \sum_k \sum_l \langle \phi^k b_{ik} \mid \phi^l b_{jl} \rangle$$

$$= \sum_{kl} \langle \phi^k \mid \phi^l \rangle b_{ik} b_{jl} = \sum_{kl} \delta_{kl} b_{ik} b_{jl}$$

$$= \sum_k b_{ik} b_{jk} \tag{3-54}$$

Equation 3-54 is almost a matrix product, but not quite. If we again invoke the definition of a transpose, we derive

$$\delta_{ij} = \sum_k b_{ik} b'_{kj} \tag{3-55}$$

which implies that

$$BB' = E \tag{3-56}$$

or

$$B^{-1} = B' \tag{3-57}$$

Linear Algebra

We have thereby proven the following results.

> **Theorem** (a) The inverse of an orthogonal matrix is the transpose of that orthogonal matrix.
> (b) The inverse of a unitary matrix is the transpose complex conjugate of that matrix.
> (c) The rows of orthogonal or unitary matrices are orthonormal vectors.
> (d) The columns of orthogonal or unitary matrices are orthonormal vectors.

Part (b) can be proved by deriving Eqs. 3–54 through 3–57 again for Hermitian vectors by supplying complex conjugation at the appropriate places. Part (c) should be clear from Eq. 3–54, and (d) is analogous. The last aspect of the structure of orthogonal and unitary transformations is a statement about the determinants of their representative matrices. We know, for orthogonal transformations, that $BB' = E$; hence,

$$|B||B'| = |E| = 1 \tag{3-58}$$

Since $|B'| = |B|$,

$$|B| = \pm 1 \tag{3-59}$$

Together with the analogous result for unitary transformations, we have proven the following theorem:

> **Theorem** The determinant of an orthogonal matrix is either $+1$ or -1; the determinant of a unitary matrix has modulus 1.

A particular, and useful, geometrical application of orthogonal transformations is rotation in two and three dimensions. We begin by considering a rotation in two dimensions. If a rotation through the angle α is carried out, the point p will be transformed into the point p'. The coordinates of p' are

$$\begin{aligned} x' &= x \cos \alpha - y \sin \alpha \\ y' &= x \sin \alpha + y \cos \alpha \end{aligned} \tag{3-60}$$

which is a set of linear equations. In matrix language, the rotational transformation may be represented by

$$R_{\text{point}} = \begin{pmatrix} \cos \alpha & -\sin \alpha \\ \sin \alpha & \cos \alpha \end{pmatrix} \tag{3-61}$$

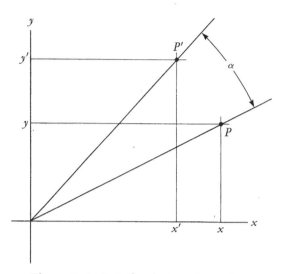

Figure 3-1 Rotation in two dimensions.

The student should notice that in this process the coordinate axes of Fig. 3-1 stay fixed but the point moves. The problem can also be described by a rotation of the coordinate axes in the opposite sense, that is, through an angle of $-\alpha$:

$$R_{\text{axis}} = \begin{pmatrix} \cos \alpha & \sin \alpha \\ -\sin \alpha & \cos \alpha \end{pmatrix} \qquad (3\text{-}62)$$

Both these points of view occur, and the important point to remember is that a rotational transformation of a point has the opposite sense of a rotation of the coordinate system.

With the matrix representative of a two-dimensional rotational transformation in mind, we may now examine rotations in three dimensions. There are many possible ways to consider a three-dimensional rotation, but all these ways have two principles in common. First, three angles are, in general, necessary to describe the rotation of a three-dimensional object. Playing with an object, preferably unsymmetrical, will convince the reader of this. Second, the customary way of describing the rotation is by the following sequence.

(a) Rotation about one of the coordinate axes by an angle ϕ.

Linear Algebra

(b) Rotation about the new position of another one of the co-ordinate axes by an angle θ.

(c) Rotation about the new position of the original axis by an angle ψ.

This sequence of rotations is called a rotation through *Eulerian angles* (after the mathematician Euler). Modern writers have made different choices for the sequence, and the student should carefully note a particular writer's convention. In agreement with a number of well-known textbooks, we shall make the following choice, illustrated in Fig. 3-2.

(a) Rotation through an angle ϕ about the z axis to yield a new $x'y'z$ axis system (matrix representative A).

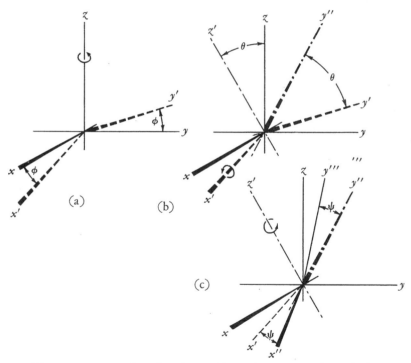

Figure 3-2 Rotation in three dimensions through Eulerian angles: (A) rotation about z by ϕ; (B) rotation about new x by θ; (C) rotation about new z by ψ.

(b) Rotation through an angle θ about the x' axis to yield a new $x'y''z'$ axis system (matrix representative B).

(c) Rotation through an angle ψ about the z' axis to yield a new, and final, $x''y'''z'$ axis system (matrix representative C).

This sequence of operations represents a rotation of the coordinate system. The matrix representatives of each separate transformation can be found by analogy with Eq. 3–62, and are

$$A = \begin{pmatrix} \cos\phi & \sin\phi & 0 \\ -\sin\phi & \cos\phi & 0 \\ 0 & 0 & 1 \end{pmatrix} \qquad (3\text{-}63a)$$

$$B = \begin{pmatrix} 1 & 0 & 0 \\ 0 & \cos\theta & \sin\theta \\ 0 & -\sin\theta & \cos\theta \end{pmatrix} \qquad (3\text{-}63b)$$

$$C = \begin{pmatrix} \cos\psi & \sin\psi & 0 \\ -\sin\psi & \cos\psi & 0 \\ 0 & 0 & 1 \end{pmatrix} \qquad (3\text{-}63c)$$

The over-all effect of these three separate rotations gives the general three-dimensional rotation matrix, R,

$$R = CBA =$$
$$\begin{pmatrix} \cos\psi\cos\phi - \cos\theta\sin\phi\sin\psi & \cos\psi\sin\phi + \cos\theta\cos\phi\sin\psi & \sin\psi\sin\theta \\ -\sin\psi\cos\phi - \cos\theta\sin\phi\cos\psi & -\sin\psi\sin\phi + \cos\theta\cos\phi\cos\psi & \cos\psi\sin\theta \\ \sin\theta\sin\phi & -\sin\theta\cos\phi & \cos\theta \end{pmatrix}$$
$$(3\text{-}64)$$

which, although complicated, plays an important role in the consideration of molecular rotation.

In conclusion of this section, and by way of offering a partial summary of the past two sections, the following lists show the interrelation between statements about transformations, matrices, determinants, and linear independence. The student should endeavor to realize the underlying unity of the subject, and to use it.

3–4 LINEAR OPERATORS

In this section we shall take a major step forward in realizing the application of the algebra of vector spaces to quantum mechanics.

Linear Algebra

Equivalence of statements: $n \times n$ matrices

Case 1. *Determinant is not zero.*

1. Linear independence of a set of vectors.
2. Row equivalence to unit matrix.
3. Matrix has an inverse.
4. Matrix has rank r equal to dimension n.
5. Linear transformation is a one-to-one mapping of n space onto n space.
6. Determinant of matrix is not zero.
7. Set of n simultaneous linear equations in n unknowns has a solution.

Case 2. *Determinant is zero.*

1. Linear dependence of a set of vectors.
2. Row equivalence to triangular matrix with at least one all-zero row.
3. Matrix has no inverse.
4. Matrix has rank r less than dimension n.
5. Linear transformation is a many-to-one mapping of n space onto r-dimensional subregion of n space.
6. Determinant of matrix is zero.
7. Set of n homogeneous simultaneous linear equations in n unknowns has a solution.

Equivalence of statements: $m \times n$ matrices with rank r

1. Linear independence of any r of the m vectors; dependence of $m - r$ of the vectors on the r independent vectors.
2. Row equivalence to triangular matrix with $m - r$ all-zero rows.
3. Linear transformation is mapping of n space onto r-dimensional subregion of m space.
4. Set of simultaneous linear equations has solution if rank of augmented matrix is also r.

Early in Chapter 1 we had mentioned the role of eigenvalue equations in quantum mechanics and seen the form which such equations take, such as Eq. 1–2. Although the description of specific operators will be left to a work on quantum mechanics per se, the general results can be set forth here. As has been our custom, we begin with a number of definitions.

> **Definition** An *operator* is a set of instructions, defined for some vector space, for changing one vector belonging to the space into another vector belonging to the space. Thus we shall write $\mathcal{A}\xi = \eta$ to mean that the result of applying a particular set of instructions, embodied in the definition of the operator $\mathcal{A}$, to the vector ξ is to form a new vector η. Operators will be notated in script letters.

We might well inquire into the difference between this definition of an operator and the definition of a transformation in the previous section. In the final analysis, there is none. The word *operator* is more often used in the context of quantum mechanics to represent particular physical quantities, whereas the word *transformation* is used to represent changes in a coordinate system. However, their definitions are identical in form, and furthermore, the definition of a linear operator is analogous to that of a linear transformation.

> **Definition** A *linear operator* $\mathcal{A}$ obeys these equations:
> (a) $\mathcal{A}(c\xi) = c\mathcal{A}\xi$, where c is a constant (perhaps complex).
> (b) $\mathcal{A}(\xi + \eta) = \mathcal{A}\xi + \mathcal{A}\eta$, where ξ and η are both vectors.

The definition of *operator* as a "set of instructions" leaves us with no way of representing operators. How may we represent linear operators? The result of operating an operator $\mathcal{A}$ on any vector can be found from the result of operating the operator $\mathcal{A}$ on the basis vectors. Suppose, for example, we know that

$$\mathcal{A}\phi^i = \sum_j A_{ij}\phi^j \qquad (3\text{-}65)$$

where the set $\{\phi^j\}$ is a complete, orthonormal basis set for the vector space in question. Any vectors may be expanded in terms of the ϕ^j vectors:

$$\xi = \sum_i c_i \phi^i \qquad \eta = \sum_j d_j \phi^j \qquad (3\text{-}66)$$

We may then use the operator definition, $\mathcal{A}\xi = \eta$, to relate the numbers $\{c_i\}$ and $\{d_i\}$:

$$\mathcal{A}\xi = \mathcal{A}\sum_i c_i \phi^i = \sum_{ij} c_i A_{ij} \phi^j = \eta = \sum_j d_j \phi^j \qquad (3\text{-}67)$$

Linear Algebra

so that

$$d_j = \sum_i A_{ij} c_i \qquad (3\text{-}68)$$

We have now only to discover what the numbers $\{A_{ij}\}$ are. This is easy if the basis set $\{\phi^i\}$ is orthonormal, for then

$$\langle \phi^j | \mathcal{a} \phi^i \rangle = \sum_k \langle \phi^j | A_{ik} \phi^k \rangle = \sum_k A_{ik} \langle \phi^j | \phi^k \rangle = A_{ij} \qquad (3\text{-}69)$$

We often see the left side of Eq. 3-69 written with an extra vertical bar, $\langle \phi^j | \mathcal{a} | \phi^i \rangle$. This extra vertical bar adds nothing new, but only serves to call attention to the operator $\mathcal{a}$ in the center. An inner product like $\langle \phi^j | \mathcal{a} | \phi^i \rangle$ is often called a *matrix element*. It will be far more convenient to use these elements as we go further. For example, Eq. 3-68 can now be written

$$d_j = \sum_i \langle \phi^j | \mathcal{a} | \phi^i \rangle c_i \qquad (3\text{-}70)$$

which has the form of a matrix product,

$$\begin{pmatrix} d_1 \\ d_2 \\ \vdots \\ d_n \end{pmatrix} = \begin{pmatrix} \langle \phi^1 | \mathcal{a} | \phi^1 \rangle & \cdots & \langle \phi^1 | \mathcal{a} | \phi^n \rangle \\ \langle \phi^2 | \mathcal{a} | \phi^1 \rangle & \cdots & \langle \phi^2 | \mathcal{a} | \phi^n \rangle \\ \vdots & & \vdots \\ \langle \phi^n | \mathcal{a} | \phi^1 \rangle & \cdots & \langle \phi^n | \mathcal{a} | \phi^n \rangle \end{pmatrix} \begin{pmatrix} c_1 \\ c_2 \\ \vdots \\ c_n \end{pmatrix} \qquad (3\text{-}71)$$

This being the case, we shall do away with the expansion coefficients A_{ij} (small capital letter), and replace them with the usual matrix elements $a_{ji} = \langle \phi^j | \mathcal{a} | \phi^i \rangle$:

$$d_j = \sum_i A_{ij} c_i = \sum_i a_{ji} c_i \qquad (3\text{-}72)$$

Here we have made use of a column matrix (an $n \times 1$ matrix) to represent a vector. An inner product may be formed from two vectors in matrix notation if the left vector is written as a $1 \times n$ row matrix and the right vector as a $n \times 1$ column matrix. Their product is, of course, a 1×1 matrix, or a scalar:

$$\langle \xi | \eta \rangle = [c_1^* \; c_2^* \; \cdots \; c_n^*] \begin{pmatrix} d_1 \\ d_2 \\ \vdots \\ d_n \end{pmatrix} = [\;\;] \qquad (3\text{-}73)$$

In our case, $\eta = \mathcal{Q}\xi$. Hence,

$$\langle \xi \mid \eta \rangle = \langle \xi \mid \mathcal{Q} \mid \xi \rangle = [c_1^* \cdots c_n^*] \begin{pmatrix} a_{11} & \cdots & a_{1n} \\ \vdots & & \vdots \\ a_{n1} & \cdots & a_{nn} \end{pmatrix} \begin{pmatrix} c_1 \\ \vdots \\ c_n \end{pmatrix} \quad (3\text{-}74)$$

Two features of importance have appeared so far. The use of a *square matrix to represent a linear operator* and the use of a *column matrix to represent a vector*. The particular matrices used were predicated on the choice of the basis set $\{\phi^i\}$, since the matrix elements were defined by $a_{ji} = \langle \phi^j \mid \mathcal{Q} \mid \phi^i \rangle$. Of course, *any* basis set would work. There are, therefore, any number of matrices which represent $\mathcal{Q}$, each matrix representative arising from a different basis. Because of this, it would, in a strict sense, be incorrect to say that the matrix A whose elements are a_{ij} is the *same* as the operator $\mathcal{Q}$; accordingly, we say that the matrix *represents* $\mathcal{Q}$, or *is a representative of* the operator $\mathcal{Q}$ in the basis $\{\phi^i\}$. Similarly, we would not be strictly correct to say that the column vector

$$\begin{pmatrix} d_1 \\ d_2 \\ \vdots \\ d_n \end{pmatrix}$$

is the *same* as η, but that this column vector *is a representative of* or *represents* η.

We should then investigate the relations between two matrix representatives of a given operator. Let A^ϕ be the representative of $\mathcal{Q}$ in the $\{\phi^i\}$ basis, and A^ψ be the representative of $\mathcal{Q}$ in the $\{\psi^i\}$ basis. The matrix elements are $a_{ij}^\phi = \langle \phi^i \mid \mathcal{Q} \mid \phi^j \rangle$ and $a_{ij}^\psi = \langle \psi^i \mid \mathcal{Q} \mid \psi^j \rangle$. Finally, let the two bases (which are orthonormal) be related by a unitary transformation,

$$\psi^i = \sum_k u_{ik} \phi^k,$$

and, in reverse,

$$\phi^i = \sum_j u_{ji}^* \psi^j = \sum_j u_{ij}'^* \psi^j.$$

The relation between the matrix elements a_{ij}^ϕ and a_{ij}^ψ is then

Linear Algebra

$$a_{ij}^{\psi} = \langle \psi^i | \mathcal{Q} | \psi^j \rangle = \sum_{k} u_{ik}^* \langle \phi^k | \mathcal{Q} | \psi^j \rangle$$

$$= \sum_{kl} u_{ik}^* u_{jl} \langle \phi^k | \mathcal{Q} | \phi^l \rangle = \sum_{kl} u_{ik}^* u_{jl} a_{kl}^{\phi}$$

$$= \sum_{kl} u'^{-1}_{ik} a_{kl}^{\phi} u'_{lj} = [U'^{-1} A^{\phi} U']_{ij} \tag{3-75}$$

The structure $A^{\psi} = U'^{-1} A^{\phi} U'$ occurs often in algebraic equations. We say that A is obtained from B by means of a *similarity transformation* if $A = S^{-1}BS$. If S is unitary, as in the present case ($S = U'$), the transformation is called a *unitary transformation;* if S is orthogonal, an *orthogonal transformation*. From this we get the following result.

> **Theorem** The matrix representative of $\mathcal{Q}$ in the ψ basis may be found by a similarity transformation of the representative of $\mathcal{Q}$ in the ϕ basis; this similarity transformation is the transpose of the transformation connecting the bases.

We have established the relations between linear operators and their matrix representatives; now it is only natural to extend to operators the features of matrix algebra, the matrix product, the inverse, and the transpose. A few further features are of importance, and are contained in the following definitions.

> **Definitions** The *commutator* of two operators $\mathcal{Q}$ and $\mathcal{B}$ is the operator $\mathcal{Q}\mathcal{B} - \mathcal{B}\mathcal{Q}$, and is designated $[\mathcal{Q}, \mathcal{B}]$.
> The *adjoint* of an operator $\mathcal{Q}$, denoted $\mathcal{Q}^{\dagger}$, is that operator whose matrix elements are related to those of $\mathcal{Q}$ by
>
> $$\langle \phi^i | \mathcal{Q}^{\dagger} | \phi^j \rangle = \langle \mathcal{Q}\phi^i | \phi^j \rangle.$$

The operator adjoint to $\mathcal{Q}$ is represented by a matrix which is related by a transposition and a complex conjugation to the representative of $\mathcal{Q}$ itself, since

$$\langle \phi^i | \mathcal{Q}^{\dagger} | \phi^j \rangle = \langle \mathcal{Q}\phi^i | \phi^j \rangle = \langle \phi^j | \mathcal{Q} | \phi^i \rangle^* \tag{3-76}$$

The dagger symbol is used to signify an adjoint.

> **Definition** The *trace* of an operator is the sum of the diagonal matrix elements of any representative of that operator: $\text{tr } \mathfrak{a} = \sum_i a_{ii}$. (In German references, *trace* is called *Spur* and is abbreviated Sp.)

As an example of some of the principles that we have been discussing, let us consider a typical operator and its representatives in some bases.

> **Definition** A *projection operator* $\mathcal{P}_\epsilon$, which gives the projection of some vector in the direction of a unit vector ϵ, is defined by $\mathcal{P}_\epsilon \xi = \langle \epsilon | \xi \rangle \epsilon$.

The projection operator gives the component of a vector in a particular direction. We may very simply solve for the eigenvalue of $\mathcal{P}_\epsilon$ using only operator methods:

$$\mathcal{P}_\epsilon^2 \xi = \mathcal{P}_\epsilon(\mathcal{P}_\epsilon \xi) = \mathcal{P}_\epsilon(\langle \epsilon | \xi \rangle \epsilon) = \langle \epsilon | \xi \rangle \mathcal{P}_\epsilon \epsilon = \langle \epsilon | \xi \rangle \epsilon \quad (3\text{-}77)$$

Hence, $\mathcal{P}_\epsilon^2 = \mathcal{P}_\epsilon$. This is expressed by saying that the operator $\mathcal{P}_\epsilon$ is *idempotent*. Then, $(\mathcal{P}_\epsilon^2 - \mathcal{P}_\epsilon) = 0 = \mathcal{P}_\epsilon(\mathcal{P}_\epsilon - 1) = 0$, and either $\mathcal{P}_\epsilon = 0$ or $\mathcal{P}_\epsilon = 1$. We already know the possible eigenvalues for $\mathcal{P}_\epsilon$—one or zero—without even setting down the matrix representative of $\mathcal{P}_\epsilon$. Now let us see what a matrix representative of $\mathcal{P}_\epsilon$ looks like.

EXAMPLE

The projection operator in a two-dimensional Euclidean vector space, with the projection direction vector $\epsilon = (1/\sqrt{2}, 1/\sqrt{2})$. Choose as the basis the conventional Cartesian basis $\phi^1 = (1, 0)$, $\phi^2 = (0, 1)$. Then, by direct evaluation,

$$p_{11}^\phi = \langle \phi^1 | \mathcal{P} | \phi^1 \rangle = \left\langle (1, 0) \middle| \left(\frac{1}{\sqrt{2}}, \frac{1}{\sqrt{2}}\right) \cdot \left(\frac{1}{\sqrt{2}}\right) \right\rangle = \frac{1}{\sqrt{2}} \cdot \frac{1}{\sqrt{2}} = \frac{1}{2}$$

$p_{12}^\phi = \frac{1}{2}$
$p_{21}^\phi = \frac{1}{2}$
$p_{22}^\phi = \frac{1}{2}$

Linear Algebra

The matrix P^ϕ is

$$\begin{pmatrix} \frac{1}{2} & \frac{1}{2} \\ \frac{1}{2} & \frac{1}{2} \end{pmatrix}$$

Consider now the basis $\psi^1 = (1/2, \sqrt{3}/2)$, $\psi^2 = (\sqrt{3}/2, -1/2)$. A calculation similar to the above gives the elements p^ψ_{ij}, and the matrix

$$P^\psi = \begin{pmatrix} \frac{1}{2} + \frac{\sqrt{3}}{4} & \frac{1}{4} \\ \frac{1}{4} & \frac{1}{2} - \frac{\sqrt{3}}{4} \end{pmatrix}$$

We must now confirm the transformation theorem, $P^\psi = U'^{-1} P^\phi U'$. Since

$$\psi^1 = \frac{1}{2}(1, 0) + \frac{\sqrt{3}}{2}(0, 1) = \frac{1}{2}\phi^1 + \frac{\sqrt{3}}{2}\phi^2$$

$$\psi^2 = \frac{\sqrt{3}}{2}(1, 0) - \frac{1}{2}(0, 1) = \frac{\sqrt{3}}{2}\phi^1 - \frac{1}{2}\phi^2$$

we have

$$U = \begin{pmatrix} \frac{1}{2} & \frac{\sqrt{3}}{2} \\ \frac{\sqrt{3}}{2} & \frac{-1}{2} \end{pmatrix} \quad U' = \begin{pmatrix} \frac{1}{2} & \frac{\sqrt{3}}{2} \\ \frac{\sqrt{3}}{2} & \frac{-1}{2} \end{pmatrix} \quad U'^{-1} = \begin{pmatrix} \frac{1}{2} & \frac{\sqrt{3}}{2} \\ \frac{\sqrt{3}}{2} & \frac{-1}{2} \end{pmatrix}$$

Then, by direct multiplication

$$\begin{pmatrix} \frac{1}{2} & \frac{\sqrt{3}}{2} \\ \frac{\sqrt{3}}{2} & \frac{-1}{2} \end{pmatrix} \begin{pmatrix} \frac{1}{2} & \frac{1}{2} \\ \frac{1}{2} & \frac{1}{2} \end{pmatrix} \begin{pmatrix} \frac{1}{2} & \frac{\sqrt{3}}{2} \\ \frac{\sqrt{3}}{2} & \frac{-1}{2} \end{pmatrix} = \begin{pmatrix} \frac{1}{2} + \frac{\sqrt{3}}{4} & \frac{1}{4} \\ \frac{1}{4} & \frac{1}{2} - \frac{\sqrt{3}}{4} \end{pmatrix} = P^\psi$$

We will soon find that in the basis $\chi^1 = (1/\sqrt{2}, 1/\sqrt{2})$, $\chi^2 = (1/\sqrt{2}, -1/\sqrt{2})$ $\mathcal{P}$ takes the form

$$\begin{pmatrix} 1 & 0 \\ 0 & 0 \end{pmatrix}$$

With these three examples we may (somewhat facetiously) illustrate the concept of matrix representation with Fig. 3-3.

We turn finally to the problem of finding the eigenvalues and eigenvectors of linear operators. Not all linear operators obey an eigenvalue equation, but, from the point of view of quantum mechanics, two classes of operators which do obey an eigenvalue equation are of great importance. These are *Hermitian operators* and *unitary operators*. We shall summarize the results first, and then offer the proofs.

> **Definition** An eigenvalue is said to be *degenerate* if the eigenvalue satisfies an eigenvalue-eigenvector equation for more than one eigenvector; the number of eigenvectors with the same eigenvalue is called the *degree of degeneracy* of the eigenvalue.
>
> **Definition** A *Hermitian operator* is a linear operator which is self-adjoint; that is, $\mathcal{H} = \mathcal{H}^\dagger$, or $h_{ij} = h_{ji}^*$. The diagonal elements of a Hermitian matrix are real.
>
> **Theorem** A Hermitian operator in an n-dimensional vector space has n *distinct eigenvectors* and n *real eigenvalues*. If the eigenvalues are nondegenerate, the eigenvectors are mutually orthogonal, and, with suitable normalization, form an orthonormal set. Even if some of the eigenvalues are degenerate, an orthonormal set can be constructed from the distinct eigenvectors.
>
> **Definition** A *unitary operator* is a linear operator whose adjoint is its inverse: $\mathcal{U}^{-1} = \mathcal{U}^\dagger$.
>
> **Theorem** A unitary operator in an n-dimensional vector space has n distinct eigenvectors (which may, as above, be made to constitute an orthonormal set), and n eigenvalues, all of which have unit modulus.

We begin by proving the Hermitian operator theorem. If $\mathcal{H}\phi^i = h_i\phi^i$, then $\langle \phi^i |\mathcal{H}| \phi^i \rangle = h_i \langle \phi^i | \phi^i \rangle$. Since $\mathcal{H}$ is Hermitian, $\langle \phi^i |\mathcal{H}| \phi^i \rangle = \langle \phi^i |\mathcal{H}| \phi^i \rangle^*$; and $h_i = h_i^*$, so h_i is real. For two different eigenvectors,

$$\mathcal{H}\phi^i = h_i\phi^i \qquad (3\text{-}78a)$$

$$\mathcal{H}\phi^j = h_j\phi^j \qquad (3\text{-}78b)$$

From Eq. 3-78a we get $\langle \phi^j |\mathcal{H}| \phi^i \rangle = h_i \langle \phi^j | \phi^i \rangle$; from Eq. 3-78b, $\langle \mathcal{H}\phi^j | \phi^i \rangle = h_j^* \langle \phi^j | \phi^i \rangle$. However, $\langle \mathcal{H}\phi^j | \phi^i \rangle = \langle \phi^i |\mathcal{H}| \phi^j \rangle^* = \langle \phi^j |\mathcal{H}| \phi^i \rangle = h_i \langle \phi^j | \phi^i \rangle$. Then, $(h_i - h_j^*)\langle \phi^j | \phi^i \rangle = 0$. Since h_j is real, if $h_i \neq h_j$, $\langle \phi^j | \phi^i \rangle = 0$ and the eigenvectors are orthogonal.

The proof of the theorem for unitary operators is very similar. If $\mathcal{U}\psi^i = u_i\psi^i$, then $\mathcal{U}^\dagger\mathcal{U}\psi^i = \psi^i = \mathcal{U}^\dagger u_i\psi^i = u_i\mathcal{U}^\dagger\psi^i$. Then $\mathcal{U}^\dagger\psi^i = (1/u_i)\psi^i$, or $\mathcal{U}^\dagger$ has the same eigenvectors as $\mathcal{U}$, but with reciprocal eigenvalues. Then $\langle \psi^i |\mathcal{U}^\dagger| \psi^i \rangle = u_i^* \langle \psi^i | \psi^i \rangle$ and $\langle \mathcal{U}^\dagger\psi^i | \psi^i \rangle =$

Linear Algebra

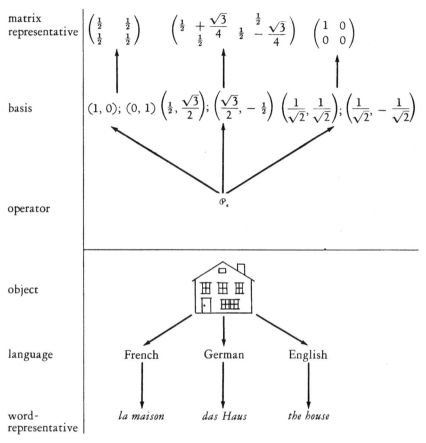

Figure 3-3 The operator $\mathcal{P}_\epsilon$, where $\epsilon = (1/\sqrt{2}, 1/\sqrt{2})$ and its matrix representatives in three bases; by analogy, the object *house*, and its word representatives in three languages.

$(1/u_i^*)\langle\psi^i | \psi^i\rangle$. However, $\langle \mathfrak{u}^\dagger\psi^i | \psi^i\rangle$ also equals $\langle\psi^i |\mathfrak{u}| \psi^i\rangle$. Hence, $1/u_i^* = u_i$, and $u_i u_i^* = 1$, or the eigenvalues u_i have modulus one. The orthogonality of the eigenvectors may be proved in a way analogous to the proof for Hermitian operators. The particular problems which arise for degenerate eigenvalues are discussed at the end of this section.

Neither of these proofs shows why Hermitian or unitary operators must obey an eigenvalue equation, and this is offered without proof. at the present. However, each of these theorems has a corollary for Euclidean vector spaces.

> **Corollary** Corresponding to the theorem for Hermitian operators in a Hermitian vector space there is the analogous theorem for *symmetric operators* ($S_{ij} = S_{ji}$) in a Euclidean vector space; corresponding to the theorem for unitary operators in a Hermitian vector space there is the analogous theorem for *orthogonal operators* ($\mathfrak{R}' = \mathfrak{R}^{-1}$) in a Euclidean vector space.

All these theorems and corollaries have dealt with the *existence* of eigenvalues and eigenvectors, but nothing so far has shown us how to *find* these eigenvalues. We can undertake a study of the solution of eigenvalue equations in general terms from two points of view, the *secular equation* and the *similarity transformation*.

Consider the eigenvalue equation, Eq. 1-2, in matrix form:

$$Q\phi = q\phi \qquad (3\text{-}79)$$

This equation is a one-line statement of a number of linear equations in the components of the vector ϕ:

$$\begin{aligned} Q_{11}\phi_1 + Q_{12}\phi_2 + \cdots + Q_{1n}\phi_n &= q\phi_1 \\ \vdots \quad \vdots \quad \vdots \quad \vdots & \\ Q_{n1}\phi_1 + Q_{n2}\phi_2 + \cdots + Q_{nn}\phi_n &= q\phi_n \end{aligned} \qquad (3\text{-}80)$$

These, in turn, may be rearranged to give a set of simultaneous linear homogeneous equations in the n unknowns ϕ_i (the components of ϕ).

$$\begin{aligned} (Q_{11} - q)\phi_1 + Q_{12}\phi_2 + \cdots + Q_{1n}\phi_n &= 0 \\ Q_{21}\phi_1 + (Q_{22} - q)\phi_2 + \cdots + Q_{2n}\phi_n &= 0 \\ \vdots \quad \vdots \quad \vdots & \\ Q_{n1}\phi_1 + Q_{n2}\phi_2 + \cdots + (Q_{nn} - q)\phi_n &= 0 \end{aligned} \qquad (3\text{-}81)$$

We have already learned that such a set of equations has a nontrivial solution only if the determinant of the coefficients vanishes. Hence, we obtain the nth degree equation for q, called the *secular equation:*

$$\begin{vmatrix} (Q_{11} - q) & Q_{12} & \cdots & Q_{1n} \\ Q_{21} & (Q_{22} - q) & \cdots & Q_{2n} \\ \vdots & \vdots & & \vdots \\ Q_{n1} & Q_{n2} & \cdots & (Q_{nn} - q) \end{vmatrix} = 0 \qquad (3\text{-}82)$$

Secular equations are commonplace in quantum mechanics. Think a minute about what Eq. 3-82 says. If we expanded Eq. 3-82 we would get a polynomial of degree n in the unknown q. Setting aside (for the moment) the practical question of solving that equation,

Linear Algebra

notice that one significant feature appears: there are n roots to the equation. The theorems we proved above stated that there would be n eigenvectors and n eigenvalues, and now we see why. It is because the eigenvalues are the solutions of an nth degree equation. Of course, as we saw, some of these roots (the eigenvalues) may be equal, but we shall defer this problem.

Very well, we now have n eigenvalues. However, the set of equations 3-81 can then be solved for the components ϕ_i for the eigenvector, *for each of the n eigenvalues*. Hence, Eq. 3-81 gives, ultimately, n eigenvectors, one eigenvector for each eigenvalue. To conclude the calculation, we shall require that the eigenvectors be normalized.

By way of recapitulation, we solve a matrix eigenvalue equation by the following steps.

1. Set up and solve the secular equation; obtain eigenvalues.

2. Use eigenvalues (now known) in eigenvalue equation to get eigenvectors.

3. Normalize eigenvectors.

We may look at the eigenvalue problem from a second point of view as well. Imagine that we know the eigenvalues and eigenvectors of a matrix:

$$\begin{aligned} Q\phi^1 &= q_1\phi^1 \\ Q\phi^2 &= q_2\phi^2 \\ &\vdots \\ Q\phi^n &= q_n\phi^n \end{aligned} \quad (3\text{-}83)$$

We could line up all the column vectors side by side and make an $n \times n$ matrix of them, such as

$$\Phi = \begin{pmatrix} \phi_1^1 & \phi_1^2 & \cdots & \phi_1^n \\ \phi_2^1 & \phi_2^2 & \cdots & \phi_2^n \\ \vdots & \vdots & & \vdots \\ \phi_n^1 & \phi_n^2 & \cdots & \phi_n^n \end{pmatrix} \quad (3\text{-}84)$$

The eigenvectors ϕ^i form the columns of the matrix Φ. The effect of operating Q on Φ is simply to generate a matrix whose columns are $q_i\phi^i$:

$$Q\Phi = \begin{pmatrix} q_1\phi_1^1 & \cdots & q_n\phi_1^n \\ q_1\phi_2^1 & \cdots & q_n\phi_2^n \\ \vdots & & \vdots \\ q_1\phi_n^1 & \cdots & q_n\phi_n^n \end{pmatrix} = \begin{pmatrix} \phi_1^1 & \cdots & \phi_1^n \\ \phi_2^1 & \cdots & \phi_2^n \\ \vdots & & \vdots \\ \phi_n^1 & & \phi_n^n \end{pmatrix} \begin{pmatrix} q_1 & & & \\ & q_2 & & \\ & & \ddots & \\ & & & q_n \end{pmatrix} \quad (3\text{-}85)$$

$$= \Phi\hat{q}$$

where we have used the symbol $\hat{q}$ to represent the matrix which is composed of the n eigenvalues q_i on the diagonal, and zeros everywhere else. Multiplying each side of Eq. 3-85 by Φ^{-1} gives

$$\Phi^{-1}Q\Phi = \hat{q} \qquad (3\text{-}86)$$

This is a statement that *the similarity transformation of Q by the matrix composed of its eigenvectors taken columnwise gives a diagonal matrix whose entries are the eigenvalues of Q.*

Hence, if we can find a way of transforming Q to a diagonal matrix, called *diagonalizing the matrix Q*, the columns of that similarity transformation are the eigenvectors and the entries in the diagonal matrix are the eigenvalues. This procedure is readily transcribed for digital computation, and often forms the basis for computer solution of eigenvalue problems. Since, as we have proven, the vectors ϕ^i are orthonormal, the transformation $\Phi^{-1}Q\Phi$ is a unitary transformation. As an example of these ideas, we find the eigenvalues of the P_ϵ matrix discussed previously.

EXAMPLE

Consider the operator $\mathcal{P}_\epsilon$, where $\epsilon = (1/\sqrt{2}, 1/\sqrt{2})$ in the usual Cartesian basis. The matrix representative of this projection operator in this basis is

$$\begin{pmatrix} \frac{1}{2} & \frac{1}{2} \\ \frac{1}{2} & \frac{1}{2} \end{pmatrix}$$

as found before. To find the eigenvalues of this operator, we solve the secular equation.

$$\begin{vmatrix} \frac{1}{2} - p & \frac{1}{2} \\ \frac{1}{2} & \frac{1}{2} - p \end{vmatrix} = 0$$

which gives the quadratic

$$\tfrac{1}{4} - p + p^2 - \tfrac{1}{4} = 0$$

which has the roots

$$p^2 - p = 0$$
$$p_1 = 1 \qquad p_2 = 0$$

These roots, which are the eigenvalues of $\mathcal{P}_\epsilon$, are just what we had predicted before from operator properties alone. We now find the eigenvector ϕ^1 corresponding to the eigenvalue $p_1 = 1$:

Linear Algebra

$$\begin{pmatrix} \frac{1}{2} & \frac{1}{2} \\ \frac{1}{2} & \frac{1}{2} \end{pmatrix} \begin{pmatrix} \phi_1{}^1 \\ \phi_2{}^1 \end{pmatrix} = \begin{pmatrix} \phi_1{}^1 \\ \phi_2{}^1 \end{pmatrix}$$

which becomes

$$\tfrac{1}{2}\phi_1{}^1 + \tfrac{1}{2}\phi_2{}^1 = \phi_1{}^1$$
$$\tfrac{1}{2}\phi_1{}^1 + \tfrac{1}{2}\phi_2{}^1 = \phi_2{}^1$$

or

$$\phi_2{}^1 = \phi_1{}^1$$
$$\phi_1{}^1 = \phi_2{}^1$$

Note that the eigenvector-component equations are identical. To fully determine the components, we require normalization, which gives $\phi_1{}^1 = \phi_2{}^1 = 1/\sqrt{2}$, or $\phi^1 = (1/\sqrt{2},\ 1/\sqrt{2})$. The eigenvector ϕ^2 corresponding to the eigenvalue $p_2 = 0$ is found by an identical process to be $\phi^2 = (1/\sqrt{2}, -1/\sqrt{2})$.

Stop and think about these eigenvectors: the eigenvector with eigenvalue one is *exactly the same* as the projection direction ϵ; that is, a vector in the projection direction projects to a vector of the same length. On the other hand, the eigenvector ϕ^2 with eigenvalue zero is perpendicular to the projection direction ϵ; a vector perpendicular to the projection direction projects to a point (zero length).

We may also confirm the similarity transformation $\Phi^{-1} P_\epsilon \Phi$ which diagonalizes P_ϵ by direct substitution:

$$\begin{pmatrix} \dfrac{1}{\sqrt{2}} & \dfrac{1}{\sqrt{2}} \\ \dfrac{1}{\sqrt{2}} & -\dfrac{1}{\sqrt{2}} \end{pmatrix} \begin{pmatrix} \dfrac{1}{2} & \dfrac{1}{2} \\ \dfrac{1}{2} & \dfrac{1}{2} \end{pmatrix} \begin{pmatrix} \dfrac{1}{\sqrt{2}} & \dfrac{1}{\sqrt{2}} \\ \dfrac{1}{\sqrt{2}} & -\dfrac{1}{\sqrt{2}} \end{pmatrix} = \begin{pmatrix} 1 & 0 \\ 0 & 0 \end{pmatrix}$$

There is another technique for finding the eigenvectors which is somewhat simpler. The components of an eigenvector obey Eq. 3-81. Each line of Eq. 3-81 looks like the expansion of a determinant by cofactors. For, if $\phi_1 = [Q - \hat{q}]_{11}$, $\phi_2 = [Q - \hat{q}]_{12}$, and so on, then the first line of Eq. 3-81 gives

$$(Q_{11} - q)\phi_1 + Q_{12}\phi_2 + \cdots + Q_{1n}\phi_n$$
$$= (Q_{11} - q)[Q - \hat{q}]_{11} + Q_{12}[Q - \hat{q}]_{12} + \cdots + Q_{1n}[Q - \hat{q}]_{1n}$$
$$= |Q - \hat{q}| = 0 \tag{3-87}$$

where the numbers Q_{ij} are the matrix elements of Q, and the numbers

q are the eigenvalues of Q. Hence the cofactor of the secular determinant for *any* row, ith column, gives a number proportional to ϕ_i.

EXAMPLE

In the case of the operator $\mathcal{P}_\epsilon$ given above, the secular determinant for the eigenvalue $p_1 = 1$ is

$$\begin{vmatrix} -\frac{1}{2} & \frac{1}{2} \\ \frac{1}{2} & -\frac{1}{2} \end{vmatrix} = 0$$

and the eigenvector ϕ^1 has components proportional to $(-\frac{1}{2}, -\frac{1}{2})$, if we choose the cofactors of the first row; ϕ^1 has components proportional to $(-\frac{1}{2}, -\frac{1}{2})$ from cofactors of the second row; each of these gives $\phi^1 = (1/\sqrt{2}, 1/\sqrt{2})$ with suitable normalization.

The student might well inquire what conditions must prevail in order that two operators have the *same* set of eigenvectors. This question is of great importance in quantum mechanics, because it tells us when two operators can both assume definite eigenvalues in the same single, stationary quantum state. Our question could equally well be phrased by asking when two matrices can be diagonalized by the same similarity transformation. The result is beautifully simple.

Theorem Two Hermitian operators can have the same set of eigenvectors (eigenfunctions) if, and only if, they commute.

Let us prove first that if $\mathcal{A}$ and $\mathcal{B}$ commute, then they have the same eigenvectors. Suppose $\mathcal{A}$ has eigenvectors ψ^i; then $\mathcal{A}\psi^i = a_i\psi^i$. However, because $\mathcal{A}$ and $\mathcal{B}$ commute, $\mathcal{A}\mathcal{B}\psi^i = \mathcal{B}\mathcal{A}\psi^i$. Then,

$$\mathcal{B}\mathcal{A}\psi^i = \mathcal{A}\mathcal{B}\psi^i = \mathcal{B}a_i\psi^i = a_i\mathcal{B}\psi^i \qquad (3\text{-}88)$$

Equation 3-88 shows that the vector $(\mathcal{B}\psi^i)$ is also an eigenvector of $\mathcal{A}$ with the eigenvalue a_i. This can only be true if $(\mathcal{B}\psi^i)$ is a multiple of ψ^i; hence,

$$\mathcal{B}\psi^i = b_i\psi^i \qquad (3\text{-}89)$$

and ψ^i is also an eigenvector of $\mathcal{B}$. The student may detect that this argument fails if the eigenvalue a_i is degenerate; such cases are considered at the end of this section.

To prove the converse—that if $\mathcal{A}$ and $\mathcal{B}$ have the same set of eigenvectors, then they commute—we begin by showing how an opera-

Linear Algebra

tor's effect may be restated in terms of projection operators. A vector ξ may be expanded in a complete orthonormal set ϕ^i:

$$\xi = \sum_i \langle \phi^i | \xi \rangle \phi^i = \sum_i \mathcal{P}_{\phi i} \xi \qquad (3\text{-}90)$$

That is, the effect of *expanding* a vector in a basis is equivalent to adding up the *projections* of that vector along the basis vectors. Suppose the eigenvectors of $\mathcal{A}$ are the set $\{\phi^i\}$. Then, because of the foregoing relation, we may write

$$\mathcal{A}\xi = \mathcal{A} \sum_i \langle \phi^i | \xi \rangle \phi^i = \sum_i \langle \phi^i | \xi \rangle a_i \phi^i = \sum_i a_i \mathcal{P}_{\phi i} \qquad (3\text{-}91)$$

which is an equation stating that the effect of operating $\mathcal{A}$ on a given vector is to generate the vector sum of the given vector projected along an eigenvector times the eigenvalue. If we represent $\mathcal{A}$ as

$$\mathcal{A} = \sum_i a_i \mathcal{P}_{\phi i} \qquad (3\text{-}92)$$

and $\mathcal{B}$, having the same set of eigenvectors $\{\phi^i\}$, as

$$\mathcal{B} = \sum_j b_j \mathcal{P}_{\phi j} \qquad (3\text{-}93)$$

then it is not difficult to show that $\mathcal{A}$ and $\mathcal{B}$ must commute, since

$$\left[\sum_i a_i \mathcal{P}_{\phi i}, \sum_j b_j \mathcal{P}_{\phi j} \right] = 0.$$

(The student should work out this commutator to his satisfaction.)

The final topic of this section is the discussion of how the properties of the eigenvalues and eigenvectors of linear operators must be modified in the case of degenerate eigenvalues. We consider first how to form an orthonormal set of eigenvectors from the eigenvectors of Hermitian (or unitary, or symmetric, or orthogonal) operators with a degenerate eigenvalue.

Suppose it happens that two eigenvectors, ξ_1 and ξ_2 have the same eigenvalue q. These vectors need not be orthogonal. Our theorems provide only that the eigenvectors corresponding to different eigenvalues are orthogonal. If $Q\xi_1 = q\xi_1$ and $Q\xi_2 = q\xi_2$, then any linear combination of ξ_1 and ξ_2 also obeys the eigenvalue equation $Q(\xi_1 + c\xi_2) = q(\xi_1 + c\xi_2)$, where c is a constant. We want the vector $\xi_1 + c\xi_2$ (*which is an eigenvector*) to be normalized and orthogonal to ξ_1. Then we will have two vectors, ξ_1 and $\xi_1 + c\xi_2$, that are orthonormal. This is precisely the problem which confronted us in the Schmidt orthogonalization procedure. Suppose ξ_1 is already normalized. Then, requiring

$$\langle \xi_1 | \xi_1 + c\xi_2 \rangle = 0 \quad (3\text{-}94)$$

gives

$$c = \frac{-\langle \xi_1 | \xi_1 \rangle}{\langle \xi_1 | \xi_2 \rangle} = -\frac{1}{\langle \xi_1 | \xi_2 \rangle} \quad (3\text{-}95)$$

This equation is analogous to Eq. 3-9; the procedure can be repeated indefinitely as indicated by the degree of degeneracy of the eigenvalue q.

Degeneracy also affects our discussion of the simultaneous eigenvectors of commuting operators. We had pointed out that if $\mathcal{A}$ and $\mathcal{B}$ are Hermitian and commute, and that if $\mathcal{A}$ has the eigenvectors $\{\psi^i\}$ and eigenvalues $\{a_i\}$, then

$$\mathcal{A}(\mathcal{B}\psi^i) = a_i(\mathcal{B}\psi^i). \quad (3\text{-}88)$$

If a_i is a nondegenerate eigenvalue, then it must be the case that $\mathcal{B}\psi^i$ is a multiple of ψ^i, as we had noted. If a_i is degenerate, however, $\mathcal{B}\psi^i$ is, in general, a linear combination of all the eigenvectors belonging to the eigenvalue a_i. Suppose these are labeled by a second index, k, running from 1 to n (we say a_i is n-fold degenerate). Then

$$\mathcal{B}\psi^i = \sum_k b_{ik}\psi^{ik} \quad (3\text{-}96)$$

The coefficients b_{ik} form an n-dimensional submatrix of the matrix representative of $\mathcal{B}$ that is not diagonal. To put it another way, if the matrix A is a diagonal, the matrix B is diagonal wherever A is nondegenerate. Where A is degenerate and B is not diagonal, we may find the eigenvalues of $\mathcal{B}$ by diagonalizing the smaller submatrix.

This section has introduced a number of ideas which are central to

Linear Algebra

a concise formulation of quantum mechanics. By way of conclusion, a summary of these concepts is presented.

1. Linear operators may be represented by matrices, once a basis set is specified. The matrix representative of the operator $\mathcal{A}$ is that matrix whose elements are $a_{ij} = \langle \phi^i | \mathcal{A} \phi^j \rangle$, where the basis $\{\phi^i\}$ is specified.

2. The effect of operating $\mathcal{A}$ on a vector ξ may be simulated by multiplying the matrix representative of $\mathcal{A}$ by a column matrix representative of ξ (both in the same basis).

3. Change of basis is effected by a similarity transformation.

4. Hermitian operators have real eigenvalues and orthonormal eigenvectors.

5. Unitary operators have eigenvalues of modulus one and orthonormal eigenvectors.

6. Eigenvalue equations may be solved by (a) setting up and solving the secular equation, substituting back into the eigenvalue equation, and normalizing; (b) solving the secular equation, using cofactors of the secular determinant to find the eigenvectors, and normalizing; or (c) finding the similarity transformation which diagonalizes the matrix.

7. Two Hermitian operators have the same set of eigenvectors if and only if they commute.

Problems

1. If the following vectors are linearly independent, form an orthonormal set from them; if they are linearly dependent, exhibit this linear dependence.
   (a) $(1, 1, 2)$, $(0, -1, 0)$, $(-1, 0, 1)$.
   (b) $(1, 2, -1, 0)$, $(0, 3, 4, 1)$, $(1, 1, 1, 1)$, $(2, 0, -4, 0)$.
   (c) $(i, 1, 2)$, $(2i + 1, -1, 3i)$, $(4, 5i, 6 + i)$.
   (d) $(0, 0, 2, 0, 0)$, $(1, 1, 1, 1, 1)$, $(1, 3, 1, 2, 2)$, $(0, 0, 1, 2, 2)$, $(1, 0, 1, 0, 1)$.

2. Are a set of mutually orthogonal vectors linearly independent? Why or why not?

3. Show that, in a vector space, $c\alpha = 0$ implies either $c = 0$ or $\alpha = 0$.

4. Form a multiplication table and compute the inner products of all vectors in Problem 1, parts (a), (b), and (c).

5. Show that matrix multiplication is associative.

6. Let
$$A = \begin{pmatrix} 1 & 2 & -1 \\ 3 & 0 & 2 \\ 4 & 5 & 0 \end{pmatrix} \quad B = \begin{pmatrix} 1 & 0 & 0 \\ 2 & 1 & 0 \\ 0 & 1 & 3 \end{pmatrix}$$

Find AB and BA. Do A and B commute? Find A^{-1} and B^{-1}. Verify that $(AB)' = B'A'$, $(AB)^{-1} = B^{-1}A^{-1}$.

7. Evaluate the determinant

$$\begin{vmatrix} 1 & -1 & 1 & -1 \\ 0 & 1 & -1 & 1 \\ 0 & 0 & 1 & -1 \\ 1 & 0 & 0 & 1 \end{vmatrix}$$

by expansion in cofactors; by direct evaluation.

8. A determinant has zeros below the main diagonal and nonzero elements on and above the main diagonal. Show that the value of this determinant is simply the product of the diagonal elements.

9. Solve the following sets of simultaneous linear equations where possible. If a solution is impossible, comment.

(a)
$$2x - 3y + 5z = 0$$
$$x - y - 2z = 2$$
$$5x \quad\quad - z = -1$$

(b)
$$2x - y + 3z - w = 0$$
$$4x - 2y - z + 3w = 0$$
$$2x - y - 4z + 4w = 0$$
$$10x - 5y - 6z + 10w = 0$$

(c)
$$2x - y + 3z = 1$$
$$4x - 2y - z = -3$$
$$2x - y - 4z = -4$$
$$10x - 5y - 6z = -10$$

(d)
$$4x + 2y + z = 11$$
$$x - y - z = -4$$
$$x + y + z = 6$$

10. What orthogonal transformation carries the Cartesian basis $(\hat{x}, \hat{y}, \hat{z})$ into the spherical basis $(\hat{r}, \hat{\theta}, \hat{\phi})$?

11. A linear transformation maps the xy plane onto the uv plane according to

$$L = \begin{pmatrix} 2 & -1 \\ -3 & 0 \end{pmatrix}$$

Find the image of the points $(1, 2)$, $(-2, 1)$, $(1, 0)$, $(0, 1)$ under L.

Linear Algebra

12. Compute the rank of each of these transformations and comment.

(a)
$$u = x + 2y - 3z$$
$$v = 2x - y + 4z$$
$$w = 3x + y + z$$

(b)
$$u = y - z$$
$$v = 3x - y + 3z$$
$$w = x + z$$

13. In a two-dimensional Hermitian space, two bases are related by

$$\psi^1 = \frac{1}{\sqrt{2}}(\phi^1 + i\phi^2) \qquad \psi^2 = \frac{1}{\sqrt{2}}(\phi^1 - i\phi^2)$$

(a) Determine the unitary matrix U which transforms the ϕ vectors into the ψ vectors.

(b) If in the ϕ representation operator a has the matrix representative

$$A^\phi = \begin{pmatrix} \cos \alpha & -\sin \alpha \\ \sin \alpha & \cos \alpha \end{pmatrix}$$

what will be its representative A^ψ in the ψ representation?

(c) Show that a unitary or orthogonal transformation always has an inverse.

14. Show that the matrix

$$A = \begin{pmatrix} a & b \\ b & b \end{pmatrix}$$

a symmetric, real matrix is transformed into the diagonal matrix

$$B = \begin{pmatrix} c & 0 \\ 0 & d \end{pmatrix}$$

by the similarity transformation $B = T^{-1}AT$, where

$$T = \begin{pmatrix} \cos \theta & -\sin \theta \\ \sin \theta & \cos \theta \end{pmatrix}$$

Derive the value of θ for which this diagonalization transformation works, and evaluate c and d.

15. Prove the following properties of Hermitian operators.
(a) Any matrix representative of a Hermitian operator has a real determinant.
(b) The inverse of a Hermitian operator is Hermitian.
(c) The product of two Hermitian operators is Hermitian if and only if they commute.

16. Prove the following properties of the *trace*.
(a) $\operatorname{tr} A^\dagger = (\operatorname{tr} A)^*$
(b) $\operatorname{tr}(aA) = a \operatorname{tr} A$
(c) $\operatorname{tr}(A + B) = \operatorname{tr} A + \operatorname{tr} B$
(d) $\operatorname{tr}(AB) = \operatorname{tr}(BA)$
(e) $\operatorname{tr}(A)$ is independent of the basis in which $\mathcal{A}$ is represented by A.

17. Show that a unitary operator may always be written in the following forms.
(a) The form $\mathcal{U} = \mathcal{A} + i\mathcal{B}$, where $\mathcal{A}$ and $\mathcal{B}$ are Hermitian, and $[\mathcal{A}, \mathcal{B}] = 0$.
(b) The form $\mathcal{U} = e^{i\mathcal{A}}$, where $\mathcal{A}$ is Hermitian.

18. Prove the following properties of orthogonal and unitary operators.
(a) The product of two orthogonal operators is orthogonal.
(b) If $\mathcal{A}$ is symmetric, and $\mathcal{U}$ orthogonal, then $\mathcal{U}^{-1}\mathcal{A}\mathcal{U}$ is symmetric.
(c) The product of two unitary operators is unitary.
(d) If $\mathcal{A}$ is Hermitian and $\mathcal{U}$ unitary, then $\mathcal{U}^{-1}\mathcal{A}\mathcal{U}$ is Hermitian.

19. Verify the matrix representing a rotation in 3D space through Eulerian angles given in Eq. 3-64.

20. What happens in the determination of eigenvectors if the eigenvalues are degenerate?

21. Find the eigenvalues and eigenvectors of the operator $\mathcal{K}$, whose matrix representative in 3D space is

$$K = \begin{pmatrix} 7 & -3 & -\sqrt{2} \\ -3 & 7 & \sqrt{2} \\ -\sqrt{2} & \sqrt{2} & 10 \end{pmatrix}$$

Check your result by applying an appropriate similarity transformation to K to diagonalize K.

22. Show that the eigenvectors of a unitary operator are orthogonal.

23. Find the eigenvalues and eigenvectors of

$$\begin{pmatrix} 1 & i & 0 \\ -i & 1 & 0 \\ 0 & 0 & 0 \end{pmatrix}$$

4

Classical Mechanics

Our purpose in this chapter is twofold. We shall first formulate classical mechanics in such a way that the connection between classical and quantum mechanics is clear.

Our second purpose is to learn to handle two mechanical problems of importance in quantum chemistry, vibrational and rotational motion of molecules. These two kinds of molecular motion lie at the center of infrared and microwave spectroscopy.

We begin with a section of review and basic definitions, proceed from there to Lagrange's and Hamilton's equations to point up the connection with quantum mechanics, and conclude with applications to molecular motion.

4-1 INTRODUCTION AND THE CONSERVATION LAWS

In considering classical mechanics, we shall freely use vector quantities. The world in which we, and our experiments, live is a three-dimensional Euclidean vector space, so that we may quickly summarize the properties of vectors in this space as follows:

(a) A vector **V** will be specified by three components, all real, usually in the Cartesian basis,

$$\mathbf{V} = (V_x, V_y, V_z) = V_x\hat{x} + V_y\hat{y} + V_z\hat{z} \qquad (4\text{-}1)$$

Vectors in this chapter are given their usual symbols, and rendered in **bold face**. Unit vectors are notated with the circumflex, or caret, or "hat" mark.

(b) The *dot product* or *scalar product* of two vectors is

$$\mathbf{V}_1 \cdot \mathbf{V}_2 = V_{1x}V_{2x} + V_{1y}V_{2y} + V_{1z}V_{2z} \qquad (4\text{-}2)$$

and is precisely the same as what we had called the *inner product* in Chapter 3. By analogy, then, the length of a vector $\mathbf{V}$ is $(\mathbf{V} \cdot \mathbf{V})^{1/2}$; two vectors are perpendicular if their dot product is zero.

In mechanics there is also defined another type of vector product.

Definition The *cross product* of two vectors is defined by

$$\begin{aligned}\mathbf{V}_1 \times \mathbf{V}_2 = &(V_{1y}V_{2z} - V_{1z}V_{2y})\hat{x} \\ &+ (V_{1z}V_{2x} - V_{1x}V_{2z})\hat{y} \\ &+ (V_{1x}V_{2y} - V_{2x}V_{1y})\hat{z}\end{aligned} \qquad (4\text{-}3)$$

Notice that a *cross product* of two vectors gives a *vector*, while the *dot product* of two vectors gives a *scalar*. The cross product of a vector with itself, that is, $\mathbf{V} \times \mathbf{V}$ is $\mathbf{0}$, if all the components of the vector commute with each other. If the components are numbers, they obviously commute, and $\mathbf{V} \times \mathbf{V} = 0$; however, if the components are operators, it will be necessary to apply the commutation rules of the operators in order to evaluate $\mathbf{V} \times \mathbf{V}$.

All our considerations in this chapter will proceed from a number of definitions, and Newton's laws. We shall, however, remain unconcerned with relativistic effects. We begin with just one particle and offer two definitions.

Definition The *linear momentum* of a particle (or simply the *momentum*) is

$$\mathbf{p} = m\mathbf{v} \qquad (4\text{-}4)$$

where m is the mass of the particle, $\mathbf{v}$ the velocity, and $\mathbf{p}$ the momentum.

Definition The *angular momentum* of a particle about a point is

$$\mathbf{l} = \mathbf{r} \times \mathbf{p} \qquad (4\text{-}5)$$

where $\mathbf{r}$ is the distance vector from the point to the particle, $\mathbf{p}$ is the (linear) momentum of the particle, and $\mathbf{l}$ the angular momentum.

Classical Mechanics

For the sake of compacting notation, we shall use the conventional dot to represent time derivatives. Thus, $\dot{\mathbf{r}} = d\mathbf{r}/dt = \mathbf{v}$; $\dot{\mathbf{v}} = d\mathbf{v}/dt = \mathbf{a}$, the acceleration, and so forth. Newton's second law is expressed as

$$\mathbf{F} = \dot{\mathbf{p}} \quad (4\text{-}6)$$

This equation, very compact, is equivalent to the familiar $\mathbf{F} = m\mathbf{a}$, if the mass is constant, since then $\dot{\mathbf{p}} = d(m\mathbf{v})/dt = m(d\mathbf{v}/dt) = m\mathbf{a}$. The simplicity of Eq. 4-6 actually includes a conservation theorem that is tantamount to Newton's first law.

> **Theorem** If the total force on a particle is zero, then the (linear) momentum of the particle does not change in time, or the momentum is *conserved*.

We may well ask whether or not a similar conservation theorem holds for angular momentum. We may show that there is such a theorem by taking the cross product of $\mathbf{r}$, the distance vector to a point of interest, with both sides of Eq. 4-6.

$$\mathbf{r} \times \mathbf{F} = \mathbf{r} \times \dot{\mathbf{p}} = \mathbf{r} \times \frac{d}{dt}(m\mathbf{v}) \quad (4\text{-}7)$$

However,

$$\frac{d}{dt} \mathbf{r} \times m\mathbf{v} = \mathbf{v} \times m\mathbf{v} + \mathbf{r} \times \frac{d}{dt}(m\mathbf{v})$$
$$= \mathbf{v} \times m\mathbf{v} + \mathbf{r} \times \dot{\mathbf{p}}$$
$$= \mathbf{r} \times \dot{\mathbf{p}} \quad (4\text{-}8)$$

since $\mathbf{v} \times \mathbf{v} = 0$ identically. Then,

$$\mathbf{r} \times \mathbf{F} = \frac{d}{dt}(\mathbf{r} \times \mathbf{p}) = \dot{\mathbf{l}} \quad (4\text{-}9)$$

The essence of Eq. 4-9 may be summarized with a definition and a theorem.

> **Definition** The *torque* about a point exerted on a particle is
> $$\mathbf{N} = \mathbf{r} \times \mathbf{F} \quad (4\text{-}10)$$
> where $\mathbf{r}$ is the distance vector from point to particle, $\mathbf{F}$ the total force, and $\mathbf{N}$ the torque.

> **Theorem** If the total torque about a point, $\mathbf{N}$, exerted on a particle is zero, then the angular momentum about that point is conserved (is unchanged in time).

The last important conservation theorem we shall discuss for a one-particle system deals with energy, and will be expressed in terms of two definitions.

> **Definition** The *work* done in moving a particle through the distance ds is
>
> $$dW = \mathbf{F} \cdot d\mathbf{s} \qquad (4\text{-}11)$$
>
> **Definition** A *conservative force* is one that may be related to a scalar *potential* by a negative derivative[1]:
>
> $$\mathbf{F} = -\frac{\partial V}{\partial x}\hat{x} - \frac{\partial V}{\partial y}\hat{y} - \frac{\partial V}{\partial z}\hat{z} = -\boldsymbol{\nabla} V = -\mathrm{grad}\, V \qquad (4\text{-}12)$$

With these definitions we can calculate the work required to move a particle from point 1 to point 2.

$$\begin{aligned} W_{12} &= \int_{(1)}^{(2)} \mathbf{F} \cdot d\mathbf{s} = \int F_x\, dx + \int F_y\, dy + \int F_z\, dz \\ &= -\int \frac{\partial V}{\partial x} dx - \int \frac{\partial V}{\partial y} dy - \int \frac{\partial V}{\partial z} dz \\ &= -V_2 - (-V_1) = V_1 - V_2 \end{aligned} \qquad (4\text{-}13)$$

On the other hand, since $\mathbf{F} = \dot{\mathbf{p}} = m\dot{\mathbf{v}}$, and $d\mathbf{s} = \mathbf{v}\, dt$, we also obtain

$$\begin{aligned} W_{12} &= \int_{(1)}^{(2)} m\dot{\mathbf{v}} \cdot \mathbf{v}\, dt = \frac{m}{2} \int \frac{d}{dt}(\mathbf{v} \cdot \mathbf{v})\, t\, d \\ &= \frac{m}{2} \int \frac{d(v^2)}{dt} dt = \frac{mv^2}{2}\bigg|_{(1)}^{(2)} \\ &= \frac{m}{2}(v_2^2 - v_1^2) \end{aligned} \qquad (4\text{-}14)$$

The quantity $mv^2/2$ is the kinetic energy, which we shall symbolize by T; hence,

---

[1] The symbol $\boldsymbol{\nabla}$, called *del* or *nabla*, is a vector differential operator: $\boldsymbol{\nabla} = +(\partial/\partial x)\hat{x} + (\partial/\partial y)\hat{y} + (\partial/\partial z)\hat{z}$. The operation of $\boldsymbol{\nabla}$ on a scalar gives a vector result, the *gradient*, abbreviated **grad**. The operation of $\boldsymbol{\nabla}$ on a vector takes two forms: $\boldsymbol{\nabla} \cdot \mathbf{a} = \mathrm{div}\, \mathbf{a}$, the *divergence* of $\mathbf{a}$, a scalar; $\boldsymbol{\nabla} \times \mathbf{a} = \mathrm{curl}\, \mathbf{a}$, a vector. The operator $\boldsymbol{\nabla} \cdot \boldsymbol{\nabla}$, or ∇^2, is called the *Laplacian*. These terms are used in a thorough examination of mechanics, electricity, and magnetism, but will be of only casual interest in this chapter.

Classical Mechanics

$$W_{12} = T_2 - T_1 = V_1 - V_2 \qquad (4\text{-}15)$$

whence

$$T_2 + V_2 = T_1 + V_1 \qquad (4\text{-}16)$$

We have proved the following theorem.

> **Theorem** Energy $(T + V)$ is conserved in motion that is the response to a conservative force.

We have discussed only one particle so far. Extending the analysis to many-particle systems is straightforward, if we agree on some additional definitions.

> **Definitions** (a) The total force on a particle i which is one of a system of particles is made up of two parts, the *external force* $\mathbf{F}_i^{\text{ext}}$ and the sum of *interparticle forces*
>
> $$\sum_{j \neq i} \mathbf{F}_{ij}^{\text{inter}}.$$
>
> Newton's third law gives $\mathbf{F}_{ij} = -\mathbf{F}_{ji}$.
> (b) The *center of mass* is located by a vector
>
> $$\mathbf{R} = \frac{\sum_i m_i \mathbf{r}_i}{\sum_i m_i} \qquad (4\text{-}17)$$
>
> where m_i is the mass of the ith particle and $\mathbf{r}_i$ the position vector of the ith particle.
> (c) The *total mass* of the system is $M = \sum_i m_i$.
> (d) The momentum of the center of mass is $\mathbf{P} = M\dot{\mathbf{R}}$.
> (e) The angular momentum of the center of mass is $\mathbf{L} = \dot{\mathbf{R}} \times \mathbf{P}$.

These five definitions allow three theorems that are the many-particle analogs of the three previous theorems, as well as three *separation theorems*. These theorems, although important, are stated without proof.

> **Theorems** (a) If the *total external force* applied to a system of particles is zero, the total linear momentum is conserved.
> (b) If the *total external torque* about a point applied to a system of particles is zero, the total angular momentum is conserved.
> (c) If both *external* and *interparticle* forces are conservative, then the energy $(T + V)$ is conserved.

The following separation theorems show how the motion of the entire system of particles may be separated into two parts.

> **Theorems** (a) The total linear momentum of a system of particles equals the linear momentum of one particle of mass M located at the center of mass.
> (b) The total angular momentum of a system of particles about a point equals the angular momentum of a single particle of mass M about the point plus the angular momentum of the particles about the center of mass.
> (c) The kinetic energy of a system of particles equals the kinetic energy of a particle of mass M moving at the velocity of the center of mass ($\dot{\mathbf{R}}$), plus the kinetic energy of the particles with respect to the center of mass.

The author acknowledges the assistance of complete textbooks on classical mechanics, particularly that of Goldstein,[2] in this and following sections. Proofs of the foregoing theorems may be found there for those interested. The six proofs are not particularly difficult, however, and appear in the problems at the end of this chapter.

In this section we have begun a study of classical mechanics by reforming Newton's laws to give conservation theorems for the momentum $\mathbf{p}$, the angular momentum $\mathbf{l}$, and the total energy, $T + V$.

4–2 GENERALIZED COORDINATES AND LAGRANGE'S EQUATIONS; HAMILTON'S EQUATIONS

In this section we shall derive forms of Newton's laws that have two particularly useful properties: these equations will be scalar

---

[2] Goldstein, H., *Classical Mechanics*, Chapter 1. Addison-Wesley, Reading, Massachusetts, 1959.

Classical Mechanics

equations, not vector equations, and these equations will be suitable for use in any coordinate system. Heretofore we had discussed both the equation of motion $\mathbf{F} = m\mathbf{a}$ and the conservation theorems—all of which are vectorial except for conservation of energy—with particular reference to a Cartesian coordinate system.

One of the tricks in solving physical or chemical problems in either classical or quantum mechanics is to choose a coordinate system judiciously, with full respect for the symmetry of the problem. For example, the motion of the earth about the sun, Kepler's problem, is difficult to describe in Cartesian coordinates, but natural in spherical coordinates. The motion of an electron about a nucleus, the quantum-mechanical hydrogen atom problem, is intractable in Cartesian coordinates, but straightforward in spherical coordinates. The advantage gained by casting the equations of motion for a system into a judiciously chosen coordinate system is great; the student will probably also appreciate the ease of handling scalar, rather than vector quantities. Of course, we wish to be absolutely *general* in examining mechanical problems. Being able to set up the equations of motion in spherical coordinates alone is not enough: we want equations of motion in a coordinate system that is *generalized*, so that any concrete coordinate system we choose is a special case of our fully generalized coordinate system.

> **Definition** The $3n$ Cartesian coordinates required to describe the mechanics of n particles may each be given by $3n$ functions of $3n$ *generalized coordinates* q_i and time:
>
> $$x_1 = x_1(q_1, q_2, \ldots, q_{3n}, t)$$
> $$y_1 = y_1(q_1, q_2, \ldots, q_{3n}, t)$$
> $$\vdots$$
> $$z_n = z_n(q_1, q_2, \ldots, q_{3n}, t)$$
>
> (4-18)

A general position vector will be notated $\mathbf{r}_i$, meaning the vector $\mathbf{r}_i = (x_i, y_i, z_i)$. Time is allowed for explicitly in Eq. 4-18 because of the possibility of moving coordinate systems, but it will not occur in the remainder of our discussion.

We begin our derivation of equations of motion of a system of particles with Newton's second law, Eq. 4-6. This is true for each particle, so that

$$\mathbf{F}_i = \dot{\mathbf{p}}_i \qquad (4\text{-}6)$$

Our derivation commences with a statement of work. This at once gives us a scalar quantity. The work done in moving particle i by an infinitesimal distance $\delta \mathbf{r}_i$ is[3]

$$\mathbf{F}_i \cdot \delta \mathbf{r}_i = \dot{\mathbf{p}}_i \cdot \delta \mathbf{r}_i \qquad (4\text{-}19)$$

We now apply the transformation equations 4-18, leaving out explicit time dependence for simplicity.

$$\mathbf{r}_i = \mathbf{r}_i(q_1, \ldots, q_{3n}) \qquad (4\text{-}20a)$$

$$\mathbf{v}_i = \dot{\mathbf{r}}_i = \sum_{k=1}^{3n} \frac{\partial \mathbf{r}_i}{\partial q_k} \frac{dq_k}{dt} = \sum_k \frac{\partial \mathbf{r}_i}{\partial q_k} \dot{q}_k \qquad (4\text{-}20b)$$

$$\delta \mathbf{r}_i = \sum_{k=1}^{3n} \frac{\partial \mathbf{r}_i}{\partial q_k} \delta q_k \qquad (4\text{-}20c)$$

The left side of Eq. 4-19 in generalized coordinates is, summed for all particles,

$$\sum_{i=1}^{n} \mathbf{F}_i \cdot \delta \mathbf{r}_i = \sum_{i=1}^{n} \sum_{j=1}^{3n} \mathbf{F}_i \cdot \frac{\partial \mathbf{r}_i}{\partial q_j} \delta q_j = \sum_{j=1}^{3n} Q_j \delta q_j \qquad (4\text{-}21)$$

A word of warning: the sums in Eq. 4-21 have different ranges, since there are n particles, but $3n$ generalized coordinates. What about the quantity Q_j that appears in Eq. 4-21? The right and left sides of Eq. 4-21 are somewhat parallel, and this prompts a definition.

Definition The *generalized force* Q_j is a scalar corresponding to the generalized coordinate q_j, and defined by

$$Q_j = \sum_{i=1}^{n} \mathbf{F}_i \cdot \frac{\partial \mathbf{r}_i}{\partial q_j} \qquad (4\text{-}22)$$

Then both sides of Eq. 4-21 have the form of (some kind of force) × (some coordinate). We have left to consider the right side of Eq. 4-19, which is more difficult. We begin with two brief statements, which are not proven here but are given as problems.

---

[3] We might be concerned about work done by forces of constraint. It is very often the case that the forces of constraint act in a direction normal to the direction of motion; hence, the work they do is nil. We shall be restricted to such cases.

$$\frac{d}{dt}\left(\frac{\partial \mathbf{r}_i}{\partial q_j}\right) = \frac{\partial \mathbf{v}_i}{\partial q_j} \tag{4-23a}$$

$$\frac{\partial \mathbf{r}_i}{\partial q_j} = \frac{\partial \mathbf{v}_i}{\partial \dot{q}_j} \tag{4-23b}$$

Then,

$$\sum_i \dot{\mathbf{p}}_i \cdot \delta \mathbf{r}_i = \sum_i m_i \ddot{\mathbf{r}}_i \cdot \delta \mathbf{r}_i = \sum_i m_i \ddot{\mathbf{r}}_i \cdot \sum_j \frac{\partial \mathbf{r}_i}{\partial q_j} \delta q_j$$

$$= \sum_{ij} \left\{ \frac{d}{dt}\left(m_i \dot{\mathbf{r}}_i \cdot \frac{\partial \mathbf{r}_i}{\partial q_j}\right) - m_i \dot{\mathbf{r}}_i \cdot \frac{d}{dt}\frac{\partial \mathbf{r}_i}{\partial q_j} \right\} \delta q_j \tag{4-24}$$

Substituting from Eq. 4-23, we obtain

$$\sum_i \dot{\mathbf{p}}_i \cdot \delta \mathbf{r}_i = \sum_{ij} \left\{ \frac{d}{dt}\left(m_i \mathbf{v}_i \cdot \frac{\partial \mathbf{v}_i}{\partial \dot{q}_j}\right) - m_i \mathbf{v}_i \cdot \frac{\partial \mathbf{v}_i}{\partial q_j} \right\} \delta q_j$$

$$= \sum_j \left\{ \frac{d}{dt}\left(\frac{\partial}{\partial \dot{q}_j}\sum_i \frac{m_i v_i^2}{2}\right) - \frac{\partial}{\partial q_j}\sum_i \frac{m_i v_i^2}{2} \right\} \delta q_j$$

$$= \sum_j \left\{ \frac{d}{dt}\frac{\partial T}{\partial \dot{q}_j} - \frac{\partial T}{\partial q_j} \right\} \delta q_j \tag{4-25}$$

since $T = \sum_i m_i v_i^2/2$. Thus we have reduced the right side of Eq. 4-19 to an expression involving scalars and generalized coordinates. Equations 4-21 and 4-25 both involve sums over displacements δq_j. Since the generalized coordinates are independent, the comparison of Eqs. 4-21 and 4-25 may be made termwise. Thus, combining the results of Eqs. 4-25 and 4-21, we establish the following theorem.

Theorem First form of *Lagrange's equations*. The equations of motion of a system of n particles in $3n$ generalized coordinates are $3n$ equations of the form

$$\frac{d}{dt}\frac{\partial T}{\partial \dot{q}_j} - \frac{\partial T}{\partial q_j} = Q_j \tag{4-26}$$

$j = 1, \cdots, 3n$, where Q_j is the generalized force of Eq. 4-22, and T is the kinetic energy.

To complete our discussion of classical mechanics in generalized coordinates, we return to the generalized force. Just as we were able to express the term $\dot{\mathbf{p}}_i \cdot \delta \mathbf{r}_i$ in terms of the kinetic energy T, we shall be able to express Q_j in terms of the potential energy V, and thereby achieve a measure of esthetic symmetry in the equations. The definition of Q_j is given in Eq. 4–22. If the forces $\mathbf{F}_i$ that appear there are conservative, they may be written as derivatives of a potential, as in Eq. 4–12. Combining Eqs. 4–12 and 4–22, we obtain

$$Q_j = \sum_i \mathbf{F}_i \cdot \frac{\partial \mathbf{r}_i}{\partial q_j} = \sum_i \left\{ F_{ix}\frac{\partial x_i}{\partial q_j} + F_{iy}\frac{\partial y_i}{\partial q_j} + F_{iz}\frac{\partial z_i}{\partial q_j} \right\}$$

$$= -\sum_i \left\{ \frac{\partial V}{\partial x_i}\frac{\partial x_i}{\partial q_j} + \frac{\partial V}{\partial y_i}\frac{\partial y_i}{\partial q_j} + \frac{\partial V}{\partial z_i}\frac{\partial z_i}{\partial q_j} \right\}$$

$$= -\frac{\partial V}{\partial q_j} \tag{4-27}$$

The first formulation of Lagrange's equations, Eq. 4–26, then becomes

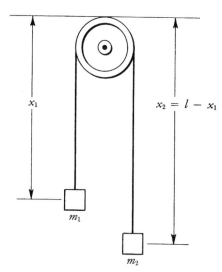

Figure 4–1 Atwood's machine, consisting of an ideal pulley and two masses.

Classical Mechanics

$$\frac{d}{dt}\frac{\partial T}{\partial \dot{q}_j} - \frac{\partial T}{\partial q_j} = -\frac{\partial V}{\partial q_j} \tag{4-28}$$

To complete the symmetrical structure of these equations we shall impose the restriction that the potential energy be independent of velocity.[4] Then, $\partial V/\partial \dot{q}_j = 0$, and Eq. 4-28 may be rewritten

$$\frac{d}{dt}\frac{\partial (T-V)}{\partial \dot{q}_j} - \frac{\partial (T-V)}{\partial q_j} = 0 \tag{4-29}$$

which leads directly to a definition and the second formulation of Lagrange's equations.

Definition The *Lagrangian* is a scalar dynamical quantity defined by

$$L = T - V \tag{4-30}$$

Theorem Second form of *Lagrange's equations*. The equations of motion of a system of n particles in $3n$ generalized coordinates are $3n$ equations of the form

$$\frac{d}{dt}\frac{\partial L}{\partial \dot{q}_j} - \frac{\partial L}{\partial q_j} = 0 \tag{4-31}$$

where L is the Lagrangian, defined in Eq. 4-30.

The Lagrangian formalism helps the recognition of constants of motion, that is, the statement of conservation theorems. In particular, we find the following theorem.

Theorem A generalized conservation theorem. If $\partial L/\partial q_j = 0$, then the dynamical quantity $\partial L/\partial \dot{q}_j$, which is a generalized momentum, is conserved.

As a brief example of the Lagrangian framework, we shall analyze the motion of the pulley-and-weight system shown in Fig. 4-1.

EXAMPLE
The system of two particles of masses m_1 and m_2, and a frictionless, weightless pulley and strings shown in Fig. 4-1 could simply be analyzed using

---

[4] This restriction is not always necessary, but it simplifies the discussion appreciably. Goldstein shows how this restriction can be circumvented.[2]

$\mathbf{F} = m\mathbf{a}$. By way of illustration, however, we work this problem in the Lagrangian framework. First, we find the potential energy (due to gravity)

$$V = -m_1 g x_1 - m_2 g(l - x_1)$$

where l is as shown in Fig. 4-1. Then, the kinetic energy is

$$T = \frac{m_1 \dot{x}_1^2}{2} + \frac{m_2 \dot{x}_2^2}{2} = \frac{m_1 + m_2}{2} \dot{x}_1^2$$

since $\dot{x}_1 = -\dot{x}_2$. The generalized coordinate system is the single coordinate x_1. The Lagrangian is

$$L = T - V = \frac{\dot{x}_1^2}{2}(m_1 + m_2) + g(m_1 - m_2)x_1 + m_2 g l$$

For Lagrange's equations we need to evaluate

$$\frac{\partial L}{\partial x_1} = (m_1 - m_2)g$$

$$\frac{\partial L}{\partial \dot{x}_1} = (m_2 + m_1)\dot{x}_1$$

There being but one coordinate, there is but one of Lagrange's equations 4-31, which is

$$\frac{d}{dt}(m_2 + m_1)\dot{x}_1 - (m_2 - m_1)g = 0$$

or

$$(m_2 + m_1)\ddot{x}_1 = (m_2 - m_1)g$$

which is precisely equivalent to the Newtonian $\mathbf{F} = m\mathbf{a}$.

The problems at the end of the chapter illustrate Lagrange's equations in situations where Newton's equation would be more clumsy.

As a consequence of expressing Newton's laws in the form of Lagrange's equations, we were able to derive a general conservation theorem: $\partial L/\partial \dot{q}_j$ is conserved if $\partial L/\partial q_j = 0$. If L is expressed in Cartesian coordinates,

$$L = T - V = \frac{m}{2}(\dot{x}^2 + \dot{y}^2 + \dot{z}^2) - V(x, y, z) \qquad (4\text{-}32)$$

then $\partial L/\partial x = F_x$, the force in the x direction, and $\partial L/\partial \dot{x} = m\dot{x} = p_x$, the momentum in the x direction. Thus, the conservation theorem in Lagrangian form reproduces the original conservation

Classical Mechanics

of momentum theorem. We might, then, to give parallel structure and equal weight to coordinate (or position) and momentum, define a generalized momentum.

> **Definition** The *generalized momentum* p_i, which is said to be *conjugate* to the generalized coordinate q_i, is defined by
>
> $$p_i = \frac{\partial L}{\partial \dot{q}_i} \quad (4\text{-}33)$$

With this definition, Lagrange's equations can be rewritten

$$\dot{p}_i = \frac{\partial L}{\partial q_i} \quad (4\text{-}34)$$

The advantage of reexpressing mechanics in terms of position and momentum, rather than in terms of position alone, is twofold. First, a striking parallel structure between position and momentum emerges which underlies much of the significance in mechanics, both classical and quantum. Second, we may rewrite the equations of motion in terms of $6n$ first-order differential equations in the $3n$ q_i's and $3n$ p_i's rather than $3n$ second-order differential equations in the q_i's alone. That is, by considering both position and momentum to be independent variables, we buy mathematical simplicity at the price of twice as many equations. To do this, that is, to write equations of motion in terms of q_i's and p_i's, we begin by defining a new dynamical quantity of mammoth significance.

> **Definition** The *Hamiltonian* for a system of n particles is defined by
>
> $$H = \sum_i p_i \dot{q}_i - L \quad (4\text{-}35)$$

Considering, as we had before, generalized coordinates with no explicit time dependence, and a potential energy independent of velocity, we may write

$$dH = \sum_i \left\{ \frac{\partial H}{\partial q_i} dq_i + \frac{\partial H}{\partial p_i} dp_i \right\} \quad (4\text{-}36)$$

where, from Eq. 4-35,

$$\frac{\partial H}{\partial q_i} = \sum p_i \frac{\partial \dot{q}_i}{\partial q_i} - \frac{\partial L}{\partial \dot{q}_i} \frac{\partial \dot{q}_i}{\partial q_i} - \frac{\partial L}{\partial q_i} \quad (4\text{-}37a)$$

$$\frac{\partial H}{\partial p_i} = \dot{q}_i \qquad (4\text{-}37\text{b})$$

Since $\partial L/\partial \dot{q}_i = p_i$, and $\partial L/\partial q_i = \dot{p}_i$, substituting into Eq. 4-36, we obtain

$$dH = \sum_i \{-\dot{p}_i dq_i + \dot{q}_i dp_i\} \qquad (4\text{-}38)$$

or

$$\dot{q}_i = \frac{\partial H}{\partial p_i} \qquad \dot{p}_i = -\frac{\partial H}{\partial q_i} \qquad (4\text{-}39)$$

Equations 4-39 are often referred to as *Hamilton's equations*. However, what are the properties of the Hamiltonian? So far, we have little more than an abstract definition, Eq. 4-35. One of the particularly important properties of H is that it is conserved, or a constant of the motion.

$$\frac{dH}{dt} = \sum_i \left(\frac{\partial H}{\partial q_i} \dot{q}_i + \frac{\partial H}{\partial p_i} \dot{p}_i \right) + \frac{\partial H}{\partial t}$$

$$= \sum_i (-\dot{p}_i \dot{q}_i + \dot{q}_i \dot{p}_i) + \frac{\partial H}{\partial t} = \frac{\partial H}{\partial t} \qquad (4\text{-}40)$$

Thus, unless H is *explicitly* time dependent, it is time independent, or a constant of the motion.

Lastly, to correlate H with more commonplace terms, we examine the kinetic energy more closely:

$$T = \frac{1}{2} \sum_i m_i v_i^2 = \frac{1}{2} \sum_i m_i \left(\sum_j \frac{\partial \mathbf{r}_i}{\partial q_j} \dot{q}_j \right)^2 \qquad (4\text{-}41)$$

if $\mathbf{r}$ is not explicitly dependent on time. Then,

$$T = \frac{1}{2} \sum_i m_i \sum_{jk} \frac{\partial \mathbf{r}_i}{\partial q_j} \cdot \frac{\partial \mathbf{r}_i}{\partial q_k} \dot{q}_j \dot{q}_k$$

$$= \frac{1}{2} \sum_{jk} t_{jk} \dot{q}_j \dot{q}_k \qquad (4\text{-}42)$$

where

Classical Mechanics

$$t_{jk} = \sum_i m_i \frac{\partial \mathbf{r}_i}{\partial q_j} \cdot \frac{\partial \mathbf{r}_i}{\partial q_k} \qquad (4\text{-}43)$$

Then,

$$\frac{\partial T}{\partial \dot{q}_i} = \sum_k t_{ik} \dot{q}_k \qquad (4\text{-}44)$$

and

$$\sum_i \dot{q}_i \frac{\partial T}{\partial \dot{q}_i} = 2T \qquad (4\text{-}45)$$

The definition in Eq. 4-34 of the generalized momentum may now be invoked. For, if, as before, the potential energy V is independent of velocity,

$$p_i = \frac{\partial L}{\partial \dot{q}_i} = \frac{\partial T}{\partial \dot{q}_i} \qquad (4\text{-}46)$$

Finally,

$$H = \sum_i p_i \dot{q}_i - L$$

$$= \sum_i \frac{\partial T}{\partial \dot{q}_i} \dot{q}_i - T + V$$

$$= 2T - T + V = T + V \qquad (4\text{-}47)$$

or, for conservative systems whose potential is independent of velocity, the *Hamiltonian is the total energy*. All these results may be expressed in a single theorem.

> **Theorem** The Hamiltonian H, for a conservative system with a velocity-independent potential energy, is a constant of the motion if it is not explicitly time dependent; it is the total energy of the system, so that, for these conditions, (total) energy is conserved. Finally, if H is independent of a (generalized) coordinate, the momentum conjugate to that coordinate is conserved; if H is independent of a momentum, the coordinate conjugate to that momentum is constant.

As a final topic in classical mechanical theory, which, together with the Hamiltonian formulation constitutes a way of bridging the gap between quantum and classical mechanics, we introduce one final definition.

Definition The *Poisson bracket* of two dynamical quantities, F and G, is defined by

$$\{F, G\} = \sum_i \left\{\frac{\partial F}{\partial q_i}\frac{\partial G}{\partial p_i} - \frac{\partial F}{\partial p_i}\frac{\partial G}{\partial q_i}\right\} \qquad (4\text{--}48)$$

One important application of Poisson brackets is in showing conservation properties, for

$$\frac{dF}{dt} = \frac{\partial F}{\partial t} + \sum_i \left\{\frac{\partial F}{\partial q_i}\dot{q}_i + \frac{\partial F}{\partial p_i}\dot{p}_i\right\}$$

$$= \frac{\partial F}{\partial t} + \sum_i \left\{\frac{\partial F}{\partial q_i}\frac{\partial H}{\partial p_i} - \frac{\partial F}{\partial p_i}\frac{\partial H}{\partial q_i}\right\}$$

$$= \frac{\partial F}{\partial t} + \{F, H\} \qquad (4\text{--}49)$$

Thus, if F is not explicitly time dependent, F is a constant of the motion (is conserved) if the Poisson bracket of F with the Hamiltonian is zero.

In this section we have developed two powerful formalisms of classical mechanics.

1. *Lagrange's equations.* $3n$ second-order differential equations in the $3n$ generalized coordinates:

$$\frac{d}{dt}\frac{\partial L}{\partial \dot{q}_i} - \frac{\partial L}{\partial q_i} = 0$$

2. *Hamilton's equations.* $6n$ first-order differential equations in $3n$ generalized coordinates and $3n$ generalized momenta:

$$\frac{\partial H}{\partial p_i} = \dot{q}_i \qquad -\frac{\partial H}{\partial q_i} = \dot{p}_i$$

These formalisms have helped us spot conservation theorems. One is particularly important: an explicitly time-independent Hamiltonian is a constant of the motion. The Hamiltonian, under certain

Classical Mechanics

common circumstances may be equated with the total energy of the system. Chapter 5 will show how Hamiltonian classical mechanics can be related to quantum mechanics.

4-3 VIBRATIONS OF A MECHANICAL SYSTEM

In this section we shall study the vibrational motion of a system of particles using the Lagrangian mechanics developed in the previous section.

There are $3n$ degrees of freedom for a system of n particles; of these, three may be accounted for in translation, and three more in rotation. Therefore, there are $3n - 6$ degrees of vibrational freedom. This brief calculation must be amended for linear molecules in which there are but two degrees of rotational freedom, and therefore $3n - 5$ degrees of vibrational freedom.

The Lagrangian is $L = T - V$. We need, therefore, to write T and V in generalized coordinates. We have, from Eqs. 4-42 and 4-43,

$$T = \tfrac{1}{2} \sum_{jk} t_{jk} \dot{q}_j \dot{q}_k \tag{4-42}$$

$$t_{jk} = \sum_i m_i \frac{\partial \mathbf{r}_i}{\partial q_j} \cdot \frac{\partial \mathbf{r}_i}{\partial q_k} \tag{4-43}$$

or, that T is a *quadratic function of the generalized velocities*. Furthermore, T may be represented by a matrix whose elements are the numbers t_{jk}.

The potential energy of a real vibrating molecule is a complicated and not completely understood function. However, the *harmonic* potential function that may be used for a simple calculation of the vibrational motion of a mechanical system is often a good approximation to the true molecular potential energy, when small displacements from equilibrium positions are considered. The mechanical system with this potential energy is called a *harmonic oscillator;* this potential energy function is identical with that which describes a system of particles obeying *Hooke's law:*

$$V = \tfrac{1}{2} \sum_{jk} v_{jk} q_j q_k \qquad (4\text{-}50)$$

or, the potential V is a *quadratic function of the generalized coordinates*. Also, V may be represented by a matrix whose elements are the numbers v_{jk}.

We may now write the Lagrangian as

$$L = T - V = \tfrac{1}{2} \sum_{jk} \{ t_{jk} \dot{q}_j \dot{q}_k - v_{jk} q_j q_k \} \qquad (4\text{-}51)$$

and, from Lagrange's equations in Eq. 4-31,

$$\frac{d}{dt} \frac{\partial L}{\partial \dot{q}_j} - \frac{\partial L}{\partial q_j} = 0 \qquad (4\text{-}31)$$

obtain

$$\frac{d}{dt} \left\{ \frac{1}{2} \sum_k t_{jk} \dot{q}_k \right\} + \frac{1}{2} \sum_k v_{jk} q_k = 0$$

$$\sum_k \{ t_{jk} \ddot{q}_k + v_{jk} q_k \} = 0 \qquad (4\text{-}52)$$

This equation bears a formal similarity, not unexpected, to that for simple harmonic motion; hence, we may try a solution of the form $q_k = \phi_k \sin \omega t$ where ϕ_k is the amplitude of the motion and ω is the vibrational frequency. This gives

$$\sum_k [v_{jk} \phi_k - \omega^2 t_{jk} \phi_k] = 0 \qquad (4\text{-}53)$$

This is just a set of simultaneous linear homogeneous equations in the variables ϕ_k. They have a solution only if the determinant of the coefficients is zero:

$$\begin{vmatrix} v_{11} - \omega^2 t_{11} & v_{12} - \omega^2 t_{12} & \cdots & v_{1N} - \omega^2 t_{1N} \\ v_{21} - \omega^2 t_{21} & v_{22} - \omega^2 t_{22} & \cdots & v_{2N} - \omega^2 t_{2N} \\ \vdots & \vdots & & \vdots \\ v_{N1} - \omega^2 t_{N1} & v_{N2} - \omega^2 t_{N2} & \cdots & v_{NN} - \omega^2 t_{NN} \end{vmatrix} = 0 \qquad (4\text{-}54)$$

This equation strongly resembles the secular equation for the eigenvalues of an operator shown in Eq. 3-82. Our present Eq. 4-54 may also be called a secular equation, in spite of the appearance of the

Classical Mechanics

eigenvalues ω^2 on and off the main diagonal and always multiplied by a matrix element of the kinetic energy matrix T. There will be $3n - 6$ (or $3n - 5$) roots ω^2, and as many values of the ϕ_k's. Similarly, the set of equations in 4-53 could be written as an eigenvector equation:

$$V\phi^i = \omega_i^2 T\phi^i \qquad (4\text{-}55)$$

That is, the effect of operating V on ϕ^i is to give a multiple, (ω_i^2), of the effect of operating T on ϕ^i.

> **Theorem** The eigenvalues, $\{\omega_i^2\}$, must be nonnegative and real.

This theorem may be proven along the same lines as given in the previous chapter.[5] It is satisfying that ω^2 is nonnegative, for then the frequency ω will be real. We may answer the eigenvalue problem for molecular vibrations in the form of another theorem.

> **Theorem** The eigenvectors, $\{\phi^i\}$, are such that a matrix Φ, whose columns are the vectors ϕ^i, simultaneously diagonalizes V and T by transformations
>
> $$\Phi'V\Phi = \Omega \qquad (4\text{-}56a)$$
>
> $$\Phi'T\Phi = E \qquad (4\text{-}56b)$$
>
> where Ω is a matrix whose diagonal elements are the possible values for ω^2, and E is the unit matrix.

The transformations here are not similarity transformations, which would have the form $\Phi^{-1}V\Phi$ and $\Phi^{-1}T\Phi$. The transformations that appear in Eqs. 4-56 are called *congruent transformations*. The eigenvalue equation 4-55 may be rewritten using the matrices Ω and Φ as

$$V\Phi = T\Phi\Omega \qquad (4\text{-}57)$$

Multiplying from the left by Φ', we have

$$\Phi'V\Phi = \Phi'T\Phi\Omega \qquad (4\text{-}58)$$

Compare this with the corresponding result for the conventional eigenvalue problem, as in Eq. 3-86, which could be rewritten

$$\Phi'Q\Phi = \Phi'\Phi\hat{q} \qquad (3\text{-}86)$$

---

[5] Goldstein, *op. cit.*, p. 323.

since Φ' is orthogonal. In Eq. 4-58, we have replaced $\Phi'\Phi$ in Eq. 3-86 with $\Phi'T\Phi$; this may be compared to the difference between Eq. 4-54 and Eq. 3-82. In fact, just as in Eq. 3-86, $\Phi'\Phi = E$, the unit matrix, we may show that $\Phi'T\Phi$ in Eq. 4-58 is also the unit matrix. For, taking the inner product of Eq. 4-55 with ϕ^j on the left,

$$\langle \phi^j |V| \phi^i \rangle = \omega_i^2 \langle \phi^j |T| \phi^i \rangle \qquad (4\text{-}59)$$

and, on the right,

$$\langle V\phi^i | \phi^j \rangle = \omega_j^2 \langle T\phi^i | \phi^j \rangle$$
$$\langle \phi^j |V| \phi^i \rangle = \omega_j^2 \langle \phi^j |T| \phi^i \rangle \qquad (4\text{-}60)$$

Comparing Eqs. 4-59 and 4-60, we see that

$$(\omega_j^2 - \omega_i^2)\langle \phi^j |T| \phi^i \rangle = 0 \qquad (4\text{-}61)$$

or, for different eigenvalues $\omega_j^2 \neq \omega_i^2$, $\langle \phi^j |T| \phi^i \rangle = 0$. Finally, by scaling the vectors $\{\phi^i\}$ appropriately, we may require $\langle \phi^j |T| \phi^j \rangle = 1$ for all ϕ^j. These statements are analogous to the orthonormality of eigenvectors in the usual eigenvalue problem.

It therefore follows that $\Phi'T\Phi = E$, as in Eq. 4-56b, and Eq. 4-58 becomes $\Phi'V\Phi = \Omega$ as in Eq. 4-56a.

This section concludes with a brief example of how this formalism may be applied to molecular vibrations. It should be pointed out at the outset that far more sophisticated methods than are considered here are available, and have been applied to the vibrations of complex molecules. However, the general features of the problem and its solution are the same.

EXAMPLE

Consider a molecule which is constrained (this is an unreal simplification) to vibrate only along a straight line. The molecule is symmetrical, as shown in Fig. 4-2.

We need first to evaluate, in the coordinate system shown (as an example—others could be chosen) the potential and kinetic energies, and then to express the potential and kinetic energies as matrices. Let us begin with the potential energy. The potential energy, in the harmonic oscillator approximation, is given by a number of terms, each of which is the square of the distention of a bond times a Hooke's law force constant:

$$V = \frac{k}{2}(x_2 - x_1 - b)^2 + \frac{k}{2}(x_3 - x_2 - b)^2$$

We are required to write the potential energy in displacement coordinates. Define the equilibrium position of the leftmost particle by x_{01}, of the center particle by x_{02}, and of the rightmost particle by x_{03}. Then define the displacement coordinates as follows:

$$q_1 = x_1 - x_{01}$$
$$q_2 = x_2 - x_{02}$$
$$q_3 = x_3 - x_{03}$$

Since

$$b = x_{02} - x_{01} = x_{03} - x_{02}$$

the potential energy is now

$$V = \frac{k}{2}(q_2 - q_1)^2 + \frac{k}{2}(q_3 - q_2)^2$$
$$= \frac{k}{2}(q_1^2 + 2q_2^2 + q_3^2 - 2q_1q_2 - 2q_2q_3)$$

We may now write down the matrix for the potential energy. The matrix elements v_{ij} are the coefficients of the term $q_i q_j$ in the potential energy expression. This means that the potential energy matrix is

$$V = \begin{pmatrix} k & -k & 0 \\ -k & 2k & -k \\ 0 & -k & k \end{pmatrix}$$

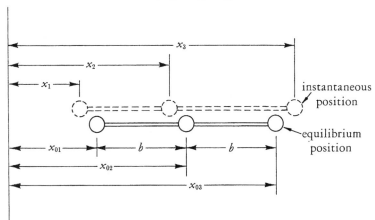

Figure 4-2 Coordinate system used to describe a linear triatomic molecule.

The kinetic energy is somewhat simpler. In the coordinate system shown in Fig. 4–2, the kinetic energy is

$$T = \frac{m}{2}(\dot{x}_1^2 + \dot{x}_3^2) + \frac{M}{2}(\dot{x}_2^2)$$

If we substitute into this equation the displacement coordinates and their time derivatives (notice that $\dot{x}_{01} = 0$, and so on), we obtain for T

$$T = \frac{m}{2}(\dot{q}_1^2 + \dot{q}_3^2) + \frac{M}{2}(\dot{q}_2^2)$$

Again taking the various terms in T (only three in this case) and placing them in a matrix where t_{ij} is the coefficient of $\dot{q}_i\dot{q}_j$ in T, we get the kinetic energy matrix

$$T = \begin{pmatrix} m & 0 & 0 \\ 0 & M & 0 \\ 0 & 0 & m \end{pmatrix}$$

The eigenvalue equation is

$$V\Phi = \omega^2 T\Phi$$

and the secular equation which must be solved is

$$|V - \omega^2 T| = 0 = \begin{vmatrix} k - \omega^2 m & -k & 0 \\ -k & 2k - \omega^2 M & -k \\ 0 & -k & k - \omega^2 m \end{vmatrix}$$

It is easy to expand the secular determinant by cofactors of its first row:

$$0 = (k - \omega^2 m)\begin{vmatrix} 2k - \omega^2 M & -k \\ -k & k - \omega^2 m \end{vmatrix} + k\begin{vmatrix} -k & -k \\ 0 & k - \omega^2 m \end{vmatrix}$$
$$= (k - \omega^2 m)[(2k - \omega^2 M)(k - \omega^2 m) - k^2] + k[-k(k - \omega^2 m)]$$
$$= (k - \omega^2 m)[(2k - \omega^2 M)(k - \omega^2 m) - k^2 - k^2]$$
$$= (k - \omega^2 m)[-\omega^2 k(M + 2m) + \omega^4 Mm]$$
$$= (k - \omega^2 m)(\omega^2)[\omega^2 Mm - k(M + 2m)]$$

This has the solutions

1. $k - \omega^2 m = 0$ $\omega^2 = \dfrac{k}{m}$ $\omega = \left(\dfrac{k}{m}\right)^{1/2}$
2. $\omega^2 = 0$ $\omega = 0$
3. $\omega^2 Mm - k(M + 2m) = 0$ $\omega = \left[\left(\dfrac{k}{m}\right)\left(1 + \dfrac{2m}{M}\right)\right]^{1/2}$

Classical Mechanics

The presence of a zero solution to the eigenvalue problem is not forbidden by the theory which has been developed. We found only that the eigenvalues, ω^2, could not be negative. If the reader went back and examined the development of the theory at that point, he would conclude that the eigenvalue $\omega = 0$ corresponds to some sort of motion where the potential energy *for that degree of freedom* is zero. Zero potential energy is not a harmonic oscillator at all; it is some sort of motion with no restoring or opposing force. In other words, it represents a free translation or free rotation.

This is what we should expect at this point, too. We have set up the problem in three generalized coordinates, but there should be only two. We have failed to eliminate the generalized coordinate which corresponds to a translation parallel to the x axis. This illustrates an important point. *Motions of the molecule that are not vibrations will, if allowed to remain in the generalized coordinate system chosen, give a zero eigenvalue for ω^2.*

We could now reformulate the problem, using a set of coordinates containing only the two which are connected with the vibration. In order to do this, we would have to formulate an equation representing our wish to forbid translation, and use it to eliminate one of the three coordinates q_1, q_2, q_3 that we had before. This equation could be chosen by requiring that the center of mass of the system remain at the origin:

$$m(x_1 + x_3) + Mx_2 = 0$$

The succeeding development would yield the same nonzero eigenvalues, but no zero eigenvalue ω^2.

We are now faced with the problem of finding the eigenvectors in our problem that are the normal modes of vibration. In order to do this, we must substitute back our (now known) eigenvalues ω^2 into the eigenvalue-eigenvector equation, and use the equation to solve for the components of the eigenvector. There will be an eigenvector for each eigenvalue, so we must carry out this process as many times as there are eigenvalues. In the case of vibrational problems, unlike the normal sort of eigenvector problem, we shall not require that the eigenvectors be normalized to one, but that $\langle \phi^i |T| \phi^i \rangle = 1$.

The eigenvalue equation reads $V\Phi = \omega^2 T\Phi$. Choosing for our first calculation the eigenvalue $\omega^2 = k/m$, we obtain

$$\begin{pmatrix} k & -k & 0 \\ -k & 2k & -k \\ 0 & -k & k \end{pmatrix} \begin{pmatrix} \phi_1^1 \\ \phi_2^1 \\ \phi_3^1 \end{pmatrix} = \frac{k}{m} \begin{pmatrix} m & 0 & 0 \\ 0 & M & 0 \\ 0 & 0 & m \end{pmatrix} \begin{pmatrix} \phi_1^1 \\ \phi_2^1 \\ \phi_3^1 \end{pmatrix}$$

Writing the matrix equation as simultaneous linear equations (homogeneous), we obtain

$$k\phi_1^1 - k\phi_2^1 \qquad\qquad = k\phi_1^1$$
$$-k\phi_1^1 + 2k\phi_2^1 - k\phi_3^1 = \frac{km}{M}\phi_2^1$$
$$\qquad\quad - k\phi_2^1 + k\phi_3^1 = k\phi_3^1$$

From either the first or last of these equations, we see that

$$\phi_2^1 = 0$$

whereas, by substituting this result into the second of these equations, we get

$$\phi_1^1 = -\phi_3^1$$

An acceptable eigenvector (before normalization) would be the vector $(a, 0, -a)$. We require the analog of normalization, namely, $\langle \phi^i | T | \phi^i \rangle = 1$. In matrix form, this is

$$(a \ \ 0 \ \ -a) \begin{pmatrix} m & 0 & 0 \\ 0 & M & 0 \\ 0 & 0 & m \end{pmatrix} \begin{pmatrix} a \\ 0 \\ -a \end{pmatrix} = (a \ \ 0 \ \ -a) \begin{pmatrix} am \\ 0 \\ -am \end{pmatrix}$$
$$= 2a^2 m = 1$$

So $a = (2m)^{-1/2}$. The normalized eigenvector is therefore $((2m)^{-1/2}, 0, -(2m)^{-1/2})$.

The same process for the eigenvalue $\omega^2 = (k/m)(1 + 2m/M)$ gives, for the unnormalized eigenvector, $(a, -2ma/M, a)$. With normalization, the eigenvector is quite complicated, $((M/\{2mM + 4m^2\})^{1/2}, -(4m/\{2M^2 + 4Mm\})^{1/2}, (M/\{2mM + 4m^2\})^{1/2})$.

The eigenvector can also be computed for the eigenvalue $\omega^2 = 0$. Unnormalized, it comes out to be (a, a, a); normalized, each component a is equal to $(2m + M)^{-1/2}$.

As a final check of this solution to the vibrational problem, we perform the transformation on the nondiagonal V matrix and on the T matrix. The result should give, for the V matrix, a diagonal matrix with the eigenvalues ω^2 on the diagonal, and, for the T matrix, the unit matrix.

For the sake of simplicity, consider a homonuclear triatomic linear molecule, such as N_3^- or I_3^-, so that $m = M$. The eigenvectors are then $((2m)^{-1/2}, 0, -(2m)^{-1/2})$, $((6m)^{-1/2}, -2(6m)^{-1/2}, (6m)^{-1/2})$, and $((3m)^{-1/2}, (3m)^{1/2}, (3m)^{-1/2})$. We may interpret these eigenvectors graphically. The eigenvector $((2m)^{-1/2}, 0, -(2m)^{-1/2})$ indicates that whenever the leftmost atom moves left, the rightmost atom moves right, and vice versa. This is called the symmetric stretch mode of vibration, and may be diagrammed as shown in Fig. 4-3a. On the other hand, the eigenvector $((6m)^{-1/2}, -2(6m)^{-1/2},$

Classical Mechanics 129

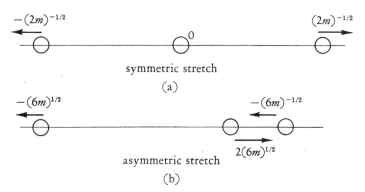

Figure 4-3 Normal modes of vibration of a linear triatomic molecule.

$(6m)^{-1/2})$ involves a motion of the end atoms (1 and 3) in the same direction, and the center atom (2) in the opposite direction. The center atom moves twice as far so as to keep the center of mass fixed. This is called the *asymmetric stretch* mode, and may be diagrammed as shown in Fig. 4-3b.

The last eigenvector, $((3m)^{-1/2}, (3m)^{-1/2}, (3m)^{-1/2})$ does not (as do the first two) keep the center of mass fixed. In fact, this eigenvector corresponds to a uniform translation of the molecule. This was the eigenvector that corresponded to the eigenfrequency of zero.

4-4 ROTATIONS OF A RIGID MECHANICAL SYSTEM

In the previous section we developed the equations of motion for a vibrating mechanical system, and showed how that analysis could be applied to molecular vibrations. In this section we develop the equations of motion for a rotating mechanical system, and show how these may be applied to molecular rotations. As was pointed out at the end of the previous section, far more elegant methods than those presented here are available; the ensuing discussion is intended only to outline the general problem.

For mechanical systems, we seek solutions to Lagrange's equations, Eq. 4-31, with the Lagrangian function defined in Eq. 4-30.

This function is particularly simple for rotations of a rigid system,[6] where $V = 0$, so $L = T$.

The kinetic energy, however, takes on a particularly simple form for rigid rotations. Since, by definition,

$$T = \tfrac{1}{2} \sum_i m_i v_i^2 \qquad (4\text{-}62)$$

we must interpret $\mathbf{v}_i$ in terms of rotations. A simple diagram will convince the student that, for circular motion,

$$\mathbf{v}_i = \boldsymbol{\omega} \times \mathbf{r}_i \qquad (4\text{-}63)$$

where $\boldsymbol{\omega}$ is the *angular velocity*. It happens that Eq. 4-63 is true in general. Using Eq. 4-63, Eq. 4-62 becomes

$$T = \tfrac{1}{2} \sum_i m_i \mathbf{v}_i \cdot (\boldsymbol{\omega} \times \mathbf{r}_i) \qquad (4\text{-}64)$$

One can show further that the order of vectors in the product of Eq. 4-64 may be permuted, giving

$$T = \frac{\boldsymbol{\omega}}{2} \cdot \sum_i m_i (\mathbf{r}_i \times \mathbf{v}_i)$$

$$= \frac{\boldsymbol{\omega}}{2} \cdot \sum_i \mathbf{l}_i = \frac{1}{2} \boldsymbol{\omega} \cdot \mathbf{L} \qquad (4\text{-}65)$$

where $\mathbf{l}_i$ is the angular momentum of the ith particle, and $\mathbf{L}$ is the angular momentum of the whole system.

In turn,

$$\mathbf{L} = \sum_i m_i (\mathbf{r}_i \times \mathbf{v}_i)$$

$$= \sum_i m_i (\mathbf{r}_i \times (\boldsymbol{\omega} \times \mathbf{r}_i))$$

$$= \sum_i m_i (\omega r_i^2 - \mathbf{r}_i (\mathbf{r}_i \cdot \boldsymbol{\omega})) \qquad (4\text{-}66)$$

---

[6] The assumption of rigidity of the molecular framework is, of course, incorrect, since the molecule does vibrate. However, this assumption simplifies the discussion and yet allows some features of interest to be manifested. Similarly, the statement $V = 0$ is not correct for the (very interesting) case of hindered rotation.

Equation 4-65 may be compactly rewritten using the definition of the moment of inertia tensor.

> **Definition** The *moment of inertia tensor* for a body, about some point, is a symmetric matrix, defined in a Cartesian coordinate system as follows. Let $q_1{}^p, q_2{}^p, q_3{}^p$ be the x, y, z coordinates of particle p. Then the i, j element of the moment of inertia is defined by
>
> $$I_{ij} = \sum_p m_p(r_p{}^2 \delta_{ij} - q_i{}^p q_j{}^p) \qquad (4\text{-}67)$$
>
> where r_p is the distance of the particle p to the particular point.

With this definition for I, we may write

$$\mathbf{L} = I\boldsymbol{\omega} \qquad (4\text{-}68)$$

and therefore

$$T = \tfrac{1}{2}\boldsymbol{\omega} \cdot I\boldsymbol{\omega} = \tfrac{1}{2}\langle \boldsymbol{\omega} | I | \boldsymbol{\omega} \rangle \qquad (4\text{-}69)$$

The question before us now is precisely how we can simply interpret the role of the moment of inertia tensor in the formula for rotational kinetic energy.

We learned in Section 4-1 that the angular momentum of a system of particles about a point may be broken up into the angular momentum of the particles about the center of mass plus the angular momentum of the center of mass. Therefore, if we focus our attention on just the rotations of a molecule, that is, if we view the molecule in a coordinate system where the center of mass is fixed, we greatly simplify the problem, for, in such a coordinate system, the angular momentum is just the angular momentum about the center of mass. Then, in Eq. 4-67, r_p may be taken to be the distance from the center of mass to the particle p.

If it is possible for us to choose a coordinate system in which I is diagonal, say the ξ, η, ζ coordinate system, then the kinetic energy would take on the particularly simple form

$$T = \frac{\boldsymbol{\omega} \cdot I\boldsymbol{\omega}}{2} = \frac{1}{2}[I_{\xi\xi}\omega_\xi{}^2 + I_{\eta\eta}\omega_\eta{}^2 + I_{\zeta\zeta}\omega_\zeta{}^2] \qquad (4\text{-}70)$$

where $I_{\xi\xi}, I_{\eta\eta}, I_{\zeta\zeta}$ are the three diagonal elements of I, and $\omega_\xi, \omega_\eta, \omega_\zeta$ are the components of the angular velocity vector.

Can I be diagonalized? Yes, since I is symmetric and real. For $i \neq j$,

$$I_{ij} = \sum_p (-m_p q_i{}^p q_j{}^p) = \sum_p (-m_p q_j{}^p q_i{}^p) = I_{ji} \qquad (4\text{-}71)$$

We may therefore appeal to the theorem of Section 3-4 dealing with the eigenvalues of Hermitian operators, for I is a Hermitian matrix. Hence the eigenvalues of I, called the *principal moments of inertia*, are real, and its eigenvectors, called the *principal axes*, are orthonormal. The similarity transformation which relates an arbitrary Cartesian axis system to the principal axis system is called the *principal axis transformation*. By diagonalizing the matrix for I, we may use at once the more simple expression for T, Eq. 4-70, in considering the energy of a rigid rotor.

It is customary to name rigid rotors according to the number of equivalent principal moments of inertia, as follows:

Moments of inertia	Name for rotor
$I_{\xi\xi} = I_{\eta\eta} = I_{\zeta\zeta}$	Spherical top
$I_{\xi\xi} = I_{\eta\eta} \neq I_{\zeta\zeta}$	Symmetric top
$I_{\xi\xi} \neq I_{\eta\eta} \neq I_{\zeta\zeta}$	Asymmetric top

EXAMPLE

Calculate the principal moments of inertia of water. If the water molecule is oriented in the coordinate system shown in Fig. 4-4, we may easily compute the elements of the moment of inertia tensor in this coordinate system. Notice that the molecule is oriented symmetrically with the center of mass at the origin of the coordinate system. We must first calculate the positions of the nuclei, given the molecular geometry. This can be done with simple trigonometry and the center-of-mass equation,

$$\sum_i m_i \mathbf{r}_i = \mathbf{0},$$

since the center of mass is at the origin. The positions of the nuclei are $\mathbf{r}(O) = (0.00, 0.07)$, $\mathbf{r}(H_a) = (-0.76, -0.55)$, and $\mathbf{r}(H_b) = (0.76, -0.55)$; $|\mathbf{r}(O)| = 0.07$, and $|\mathbf{r}(H_a)| = |\mathbf{r}(H_b)| = 0.94$. Using Eq. 4-67, we may

Classical Mechanics

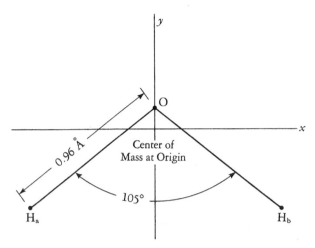

Figure 4-4 Principal axes of water molecule.

now compute the moment of inertia tensor for water about its center of mass. For example,

$$I_{xx} = \sum_p m_p(r_p^2 - x_p^2)$$

$$= [16(0.0049 - 0.0000) + 1(0.881 - 0.578)$$
$$+ 1(0.881 - 0.578)](M)(10^{-16}) = 1.21 \times 10^{-40} \text{ g cm}^2$$

where M is the mass of the proton, 1.67×10^{-24} g, and $1 \text{ Å}^2 = 10^{-16} \text{ cm}^2$. Likewise,

$$I_{xy} = -\sum_p m_p x_p y_p$$

$$= -[(16)(0)(0.07) - (1)(-0.76)(-0.55)$$
$$- (1)(0.76)(-0.55)](M)(10^{-16}) = 0$$

In a similar fashion, we calculate the tensor I to be

$$I = \begin{pmatrix} 1.21 \times 10^{-40} & 0 & 0 \\ 0 & 1.93 \times 10^{-40} & 0 \\ 0 & 0 & 3.08 \times 10^{-40} \end{pmatrix} \text{ g cm}^2$$

We find the principal moments of inertia at once, since the tensor is diagonal. Furthermore, we find that the x, y, and z axes of Fig. 4-4 are the principal axes.

This classification of molecules, and the methods of matrix algebra devoted to finding the principal moments of inertia have motivated the preceding section.

Problems

1. Prove the properties of the cross product mentioned beneath Eq. 4-3.
2. Prove the three conservation theorems for systems of particles suggested in Section 4-1.
3. Prove the three separation theorems for systems of particles suggested in Section 4-1.
4. Two particles of mass m are joined by a rigid, weightless rod of length l; the center of the rod is constrained to move frictionlessly on a circular wire loop of radius a. Choose a suitably simple set of generalized coordinates and express T in this set.
5. Obtain the Lagrange equations for a pendulum which may move in three dimensions: a particle of mass m suspended under the influence of gravity by a rigid, weightless rod.
6. Two particles of unequal mass are connected by a (weightless, inelastic) string. One particle slips over the side of a table while the other rests on the (frictionless) table top. Describe the motion using Lagrangian mechanics.
7. A ball of mass m is attached to the center of a smooth flat table by means of a spring. The force constant of the spring is k, and its equilibrium length is r_0. Find the equations of motion for the system, and comment.
8. (a) Derive Eq. 4-23a; (b) derive Eq. 4-23b.
9. A particle is held inside a flat rectangular box by four springs, two each of two different force constants. The particle is given a small displacement from equilibrium. Describe its motion.
10. Derive Eq. 4-53.
11. Reformulate and work out the normal modes of the linear triatomic molecule in the center-of-mass coordinate system as hinted at in the example of Section 4-3.
12. Find the normal modes and frequencies of vibration of water.
13. Find the normal modes and frequencies of vibration of carbon dioxide.
14. Derive Eqs. 4-63, 4-65, and 4-66.
15. Prove Euler's theorem: The most general displacement of a rigid body with one point fixed is a rotation about some axis.
16. Work through the example given in Section 4-4 to calculate the moment of inertia tensor for water in a coordinate system where the oxygen atom is at the origin and one O—H bond lies in the x axis, rather than in

the coordinate system of Fig. 4-4. Compare the results from your calculation to those of the example.

17. The molecules CCl_4, NCl_3, OCl_2, and FCl are all chlorides of first-row elements. Compute the principal moments of inertia of each, and classify each as a particular kind of "top." Sketch each, and show the principal axes. You will have to look up the structure of these molecules.

18. Show that tetrahedral and cubic molecules are spherical tops.

5

Conclusion

The four preceding chapters of this book have been designed to prepare the stage for quantum mechanics, and especially for molecular spectroscopy. To finish this stage setting—in fact, to raise the curtain a bit—we touch briefly on two points. The first is the connection between classical and quantum mechanics; the second is the synthesis of the wave-mechanical and matrix-mechanical points of view.

5-1 THE BRIDGE BETWEEN CLASSICAL AND QUANTUM MECHANICS

To introduce quantum mechanics properly, one should become aware of the thrilling experiments performed around the end of the nineteenth and beginning of the twentieth century which begged a theoretical explanation not afforded by Newtonian mechanics and Maxwellian electromagnetism. To review this background would carry us away from the purpose of this book. This section is accordingly confined to a comparison of classical and quantum mechanics at the level of elementary theory, and the experimental basis

Conclusion

for quantum mechanics is left for the enjoyment of the reader elsewhere.

Traditionally, quantum mechanics courses begin with a review of the evidence for the failure of classical mechanics. They approach quantum mechanics most often through the (quite natural) point of view of wave-particle duality, exemplified by the work of de Broglie. It should be understood at the outset that the conventional admixture of classical wave equations with the de Broglie momentum-wavelength relation does not constitute a derivation of the Schrödinger equation, the basic equation from which wave mechanics proceeds. The Schrödinger equation may be taken only as a postulate from which an outlook on the physical world can be built, for better or for worse. As such, this outlook has been enormously successful in describing submicroscopic matter, requiring assistance only in the directions of relativity and field theory. Wave mechanics, however, is not *derived;* it is *postulated.* An alternative approach to quantum mechanics—one which is at the same time more abstract and more fundamental—is to establish a connection between operators and corresponding dynamical quantities. In this direction, one can propose a simple and self-consistent set of postulates for quantum mechanics which are at once elegant, abstract, and basic.[1] Again, it should be understood at the outset that the approach to be presented here is not *derived,* but *postulated.* As such, it has neither more nor less physical content than Schrödinger's mechanics. The author hopes only that an alternative formulation of quantum mechanics, using the terminology of Chapters 3 and 4, rather than the (perhaps less familiar) terminology of wave motion will at least cast new light on the comparison of quantum to classical mechanics, and perhaps even make bridging the gap between the two physics less a leap of faith and more one of reason.

It was Bohr[2] who realized that, because classical mechanics was "right" for macroscopic systems, the introduction of quantum mechanics could not totally eradicate the classical theory which had

---

[1] See, for example, the excellent treatment in Dicke, R. H., and Wittke, J. P., *Introduction to Quantum Mechanics,* Chapter 6. Addison-Wesley, Reading, Massachusetts, 1960.

[2] Bohr, N., "The Quantum Postulate and the Recent Development of Atomic Theory," *Nature* **121,** 580 (1928).

gone before. Somehow, quantum mechanics must contain classical mechanics as a limiting case.[3] This statement itself may be thought of as the *correspondence principle*. How may we formulate the nature of the limit in which quantum mechanics yields classical mechanics? This question can be answered at many levels of sophistication. We offer here only one way of pointing out the classical limit of quantum mechanics and also the correspondence between them; there are other methods—some very elegant—for demonstrating this correspondence.

Following Dicke and Wittke,[4] for example, we might begin a postulatory formulation of quantum mechanics by saying that to every physical observable there corresponds an operator; for physically measurable quantities the operator must be Hermitian; the result of measuring the physical quantity is one of the eigenvalues of the operator which corresponds to that quantity.

In the limit of classical mechanics, the operator nature of physical quantities is lost; or, to be more precise, the operator for a physical quantity Q, denoted $\mathcal{Q}$, must be defined by the instructions "multiply the operand function by Q," that is, $\mathcal{Q}f \rightarrow Qf$. If this is so, then, in the classical limit, all operators corresponding to physical quantities commute. In the situations wherein quantum mechanics alone "rightly" describes the problem, all operators do not commute. Hence the nature of the classical limit of quantum mechanics must be

$$[\mathcal{A}, \mathcal{B}] = \text{(small number)(operator corresponding to some function of } A, B) \quad (5\text{-}1)$$

In the development of quantum theory, one "small number" appears time and time again. This number is Planck's constant. In fact, $i\hbar$ is just the right number for Eq. 5-1, as was verified by eventually comparing theory to experiment. We have now

$$[\mathcal{A}, \mathcal{B}] = i\hbar(\text{operator corresponding to some function of } A, B) \quad (5\text{-}2)$$

Our search for a correspondence between classical and quantum me-

---

[3] Landau, L. D., and Lifschitz, E. M., *Quantum Mechanics: Non-relativistic Theory*, p. 19. Translated from Russian by J. B. Sykes and J. S. Bell (Volume 3 in *Course in Theoretical Physics*). Addison-Wesley, Reading, Massachusetts, 1958.

[4] Dicke and Wittke, *op. cit.*, pp. 91 ff.

Conclusion

chanics (which is here obviously not derived) is almost over. For reasons of dimensions, the "function of A, B" which we have yet to supply, must have the dimensions of $(AB/\text{erg-sec})$, since $i\hbar$ has units of erg-sec. Such a function is the Poisson bracket of A and B, whence

$$[\mathcal{A}, \mathcal{B}] = i\hbar(\text{operator corresponding to } \{A, B\}) \quad (5\text{-}3)$$

which is precisely the seventh postulate of Dicke and Wittke.[5] Equation 5-3 is a "prescription" or a "helpful guide" rather than a derivation. Proceeding from this statement of correspondence and limit between classical and quantum mechanics, we may derive (and now *derive* is the right word) some consequences of importance.

First, compare the Poisson bracket of a quantity with the classical Hamiltonian to the commutator of the corresponding operator with the Hamiltonian operator.

$$df/dt = \{f, H\} + \partial f/\partial t \quad (4\text{-}48)$$

Again appealing to the principle that in the classical limit, the operator $\mathcal{F}$ must behave just like the function f,

$$\frac{d\mathcal{F}}{dt} = -\frac{i}{\hbar}[\mathcal{F}, \mathcal{H}] + \frac{\partial \mathcal{F}}{\partial t} = \frac{i}{\hbar}[\mathcal{H}, \mathcal{F}] + \frac{\partial \mathcal{F}}{\partial t} \quad (5\text{-}4)$$

which is *Heisenberg's equation of motion for the operator* $\mathcal{F}$. Heisenberg's point of view, which placed emphasis on linear operators and their matrix representatives, put all the time dependence for a quantity into its corresponding operator through Eq. 5-4. This equation allows us to state the correspondence principle in terms of a conservation theorem: *Quantities which are constants of the motion in classical mechanics remain constants of the motion in quantum mechanics.* In particular, if the classical Hamiltonian is not explicitly time dependent, then energy is conserved:

$$\frac{dH}{dt} = \{H, H\} + \frac{\partial H}{\partial t} = \{H, H\} = 0 \quad (5\text{-}5a)$$

This has the quantum-mechanical analog

$$\frac{d\mathcal{H}}{dt} = \frac{i}{\hbar}[\mathcal{H}, \mathcal{H}] + \frac{\partial \mathcal{H}}{\partial t} = \frac{i}{\hbar}[\mathcal{H}, \mathcal{H}] = 0 \quad (5\text{-}5b)$$

---

[5] Dicke and Wittke, *op. cit.*, p. 102.

An alternative point of view,[6] that taken by Schrödinger, is to leave operators implicitly independent of time, and to make their eigenfunctions time dependent. To do this, begin with the equation of motion of an eigenvalue f of the operator whose eigenfunction is ϕ. Then

$$\frac{d\mathcal{F}}{dt} = \frac{d}{dt}\langle \phi | \mathcal{F} | \phi \rangle = \langle \frac{d\phi}{dt} | \mathcal{F} | \phi \rangle + \langle \phi | \mathcal{F} | \frac{d\phi}{dt} \rangle \quad (5\text{-}6)$$

From Eq. 5-4,

$$\frac{d\mathcal{F}}{dt} = \langle \phi | \frac{i}{\hbar}[\mathcal{H}, \mathcal{F}] | \phi \rangle = \langle \phi | \frac{i}{\hbar}\mathcal{H}\mathcal{F} - \frac{i}{\hbar}\mathcal{F}\mathcal{H} | \phi \rangle$$
$$= \langle -\frac{i}{\hbar}\mathcal{H}\phi | \mathcal{F} | \phi \rangle + \langle \phi | \mathcal{F} | -\frac{i}{\hbar}\mathcal{H}\phi \rangle \quad (5\text{-}7)$$

whence, in order to have agreement between Eqs. 5-6 and 5-7, we require

$$\mathcal{H}\phi = i\hbar \frac{d\phi}{dt} \quad (5\text{-}8)$$

which is Schrödinger's equation.

Thus we have seen that a neat correspondence can be established between classical and quantum mechanics which introduces both Heisenberg's and Schrödinger's equations of motion, and concomitant conservation laws.

5-2 THE SYNTHESIS OF MATRIX AND WAVE MECHANICS

We have mentioned the two points of view of eigenvalue-eigenvector problems. In the first, the operator is a differential operator and the eigenvalue equation is a differential equation. In the second, the operator is represented by a matrix and the eigenvalue equation by a set of simultaneous linear homogeneous equations. The syn-

---

[6] The words *point of view* have been used here in describing the way in which equations of motion are cast in quantum mechanics. Others use the word *representation* for this concept; thus, *Heisenberg representation* or *Schrödinger representation*. We, however, have used *representation* to mean something quite different, and therefore avoid using it here.

Conclusion

thesis of these two points of view is found in the concept of Hilbert space.[7]

We need to show how the algebraic methods of Chapter 3 remain applicable for problems with a continuous variable (as many quantum-mechanical problems have). If we define a *Hilbert space* as a vector space spanned by normalizable functions with a defined inner product (Eq. 3-1), then we may establish a synthesis of matrix and wave mechanics. The concept of Hilbert space allows a vector space with infinite dimension. This infinite number may be discrete and denumerable, as in the index n in the basis set $\{H_n(x)\}$, $n = 0, 1, \ldots$. Or, this infinite number may be continuous and non-denumerable, as the number of points on the line.

Further discussion of Hilbert space would carry us too far afield; the author hopes only to indicate the nature of the similarity between an eigenfunction and an eigenvector. In Hilbert space, the matrix representative of an operator might be of infinite dimension; an eigenfunction would be representable by an infinitely long column vector. For example, consider the operator for position, $\mathfrak{X}$, which has a continuous range, or *spectrum* of eigenvalues, say $\{x'\} : \mathfrak{X} \mid x'\rangle = x' \mid x'\rangle$. Any eigenvector $\mid \xi\rangle$ in the Hilbert space may be expanded in terms of the eigenvectors of $\mathfrak{X}$ (since they comprise a complete, orthonormal set).

$$\mid \xi\rangle = \sum \mid x'\rangle\langle x' \mid \xi\rangle = \int \mid x'\rangle\langle x' \mid \xi\rangle \, dx' \qquad (5\text{-}9)$$

Since the Hilbert space is of infinite dimension, we replace the sum by an integral over the continuous variable x'. The coefficients $\langle x' \mid \xi\rangle$ are (complex) functions which make up the representative of the vector $\mid \xi\rangle$ in the basis $\mid x'\rangle$. We may call this collection of complex functions the *wave function:* $\Psi_\xi(x') = \langle x' \mid \xi\rangle$. The inner product of two Hilbert space vectors, $\langle \xi \mid \eta\rangle$, provides the final link to the theory of Chapter 2:

$$\begin{aligned}\langle \xi \mid \eta\rangle &= \int dx' \int dx'' \langle \xi \mid x'\rangle\langle x' \mid x''\rangle\langle x'' \mid \eta\rangle \\ &= \int dx' \langle \xi \mid x'\rangle\langle x' \mid \eta\rangle \\ &= \int dx' \, \Psi_\xi^*(x')\Psi_\eta^*(x') \end{aligned} \qquad (5\text{-}10)$$

which is the same as Eq. 2-1.

---

[7] For a somewhat more detailed discussion, see Jackson, J. D., *Mathematics for Quantum Mechanics*, Sections 5-10 and 5-11. W. A. Benjamin, New York, 1962.

Thus we conclude our foray into the mathematics and physics which are relevant to quantum chemistry. The author's hope is that the student's mastery of the principles and applications of quantum mechanics will not be needlessly fettered by a simultaneous strain to grasp the mathematical language in which the subject is couched.

Problems

1. Show that Eq. 5-3 is dimensionally correct.
2. Using the correspondence principle as expressed in Eq. 5-3, find the commutation relations between Cartesian coordinates and Cartesian linear momenta.

Appendix

Mathematical Background and Bibliography

The purpose of these brief appendices is to summarize some formulas of importance that may be unfamiliar to students with two years of college mathematics, and yet are important to the development of the text.

A-1 COMPLEX NUMBERS

(A) ARITHMETIC OF COMPLEX NUMBERS
 1. Definition: $i = \sqrt{-1}$.
 2. Addition: If $z = x + iy$, $z' = x' + iy'$, then $(z + z') = (x + x') + i(y + y')$.
 3. Modulus: $|z| = (x^2 + y^2)^{1/2}$.
 4. Multiplication: $zz' = (xx' - yy') + i(xy' + yx')$.
 5. Division: $z/z' = \{(xx' + yy') + i(xy' - x'y)\}/\{x'^2 + y'^2\}$.
 6. Conjugation: $z^* = x - iy$.

(B) ALGEBRAIC THEOREMS
 1. Modulus of sum of two numbers is less than or equal to sum of moduli: $|z_1 + z_2| \leq |z_1| + |z_2|$.
 2. Decomposition into modulus and argument: $z = x + iy = r(\cos\theta + i\sin\theta)$, where $r = |z|$, $\theta = \tan^{-1}(y/x)$.

3. De Moivre's theorem: If $z = r(\cos\theta + i\sin\theta)$, then $z^n = r^n(\cos n\theta + i\sin n\theta)$.

(C) SPECIAL FUNCTIONS

1. Exponential. Defined by the power series, $\exp[z] = 1 + z + z^2/2! + z^3/3! + \cdots$, the exponential function has all the same properties as the exponential of a real variable. In particular,

$$e^z e^{z'} = e^{z+z'} \qquad e^{iy} = \sum_n (iy)^n/n! = \cos y + i\sin y.$$

Hence, $e^z = e^x(\cos y + i\sin y)$.

2. Trigonometric. Defined by power series, these functions are apparently identical in all respects to their real counterparts. Notice, from paragraph 1 above, that

$$\cos z = \frac{e^{iz} + e^{-iz}}{2} \qquad \sin z = \frac{e^{iz} - e^{-iz}}{2i}.$$

A-2 CALCULUS: PARTIAL DERIVATIVES

(A) DEFINITIONS

1. Total derivative:

$$\left.\frac{dy}{dx}\right|_{x_0} = \lim_{\Delta x \to 0} \frac{\Delta y}{\Delta x} = \lim_{\Delta x \to 0} \frac{y(x_0 + \Delta x) - y(x_0)}{\Delta x}$$

2. Partial derivative:

$$\left.\frac{\partial z}{\partial x}\right|_{x_0, y_0} = \lim_{\Delta x \to 0} \frac{z(x_0 + \Delta x, y_0) - z(x_0, y_0)}{\Delta x}$$

(B) THEOREMS

1. Inversion.

$$\left.\frac{\partial z}{\partial y}\right|_x = \left(\left.\frac{\partial y}{\partial z}\right|_x\right)^{-1}$$

2. Chain rule. Example: suppose $w = w(u, v)$, $u = u(x, y)$, $v = v(x)$. Ultimately, w is dependent on x and y. Evaluate this dependence by tracing through the intermediate functions u, v:

$$\frac{\partial w}{\partial x} = \frac{\partial w}{\partial u}\frac{\partial u}{\partial x} + \frac{\partial w}{\partial v}\frac{dv}{dx}$$

$$\frac{\partial w}{\partial y} = \frac{\partial w}{\partial u}\frac{\partial u}{\partial y}$$

Appendix

3. Implicit differentiation. Example: If u, v, x, y are interrelated by

$$F(u, v; x, y) = 0 \qquad G(u, v; x, y) = 0$$

then solve the two simultaneous linear equations

$$0 = \frac{\partial F}{\partial x} + \frac{\partial F}{\partial u}\frac{\partial u}{\partial x} + \frac{\partial F}{\partial v}\frac{\partial v}{\partial x}$$

$$0 = \frac{\partial G}{\partial x} + \frac{\partial G}{\partial u}\frac{\partial u}{\partial x} + \frac{\partial G}{\partial v}\frac{\partial v}{\partial x}$$

for $\partial u/\partial x$ and $\partial v/\partial x$; similarly for $\partial u/\partial y$ and $\partial v/\partial y$. Application of the theory of Chapter 3 will simplify such a calculation by defining the *Jacobian*,

$$J(F, G; x, y) = \begin{vmatrix} \dfrac{\partial F}{\partial x} & \dfrac{\partial F}{\partial y} \\ \dfrac{\partial G}{\partial x} & \dfrac{\partial G}{\partial y} \end{vmatrix}$$

Then, $\partial u/\partial x = -J(F, G; x, v)/J(F, G; u, v)$, and so forth.

4. Triple product.

$$\left(\frac{\partial x}{\partial z}\bigg|_y\right)\left(\frac{\partial z}{\partial y}\bigg|_x\right)\left(\frac{\partial y}{\partial x}\bigg|_z\right) = -1.$$

A-3 BIBLIOGRAPHY

The following bibliography is not intended to be complete, but rather a survey of works that the author has found useful.

(A) MATHEMATICS FOR PHYSICS AND CHEMISTRY

Chisholm, J. S. R., and Morris, R. M., *Mathematical Methods in Physics.* W. B. Saunders, Philadelphia, 1965.

Courant, R., and Hilbert, D., *Methods of Mathematical Physics*, Vol. I., 1st English ed. Interscience, New York, 1953.

Irving, J., and Mullineux, N., *Mathematics in Physics and Engineering.* Academic Press, New York, 1959.

Jackson, J. D., *Mathematics for Quantum Mechanics.* W. A. Benjamin, New York, 1962.

Margenau, H., and Murphy, G. M., *The Mathematics of Physics and Chemistry.* Van Nostrand, Princeton, 1956.

Matthews, J., and Walker, R. L., *Mathematical Methods of Physics*. W. A. Benjamin, New York, 1964.

Sokolnikoff, I. S., and Redheffer, R. M., *Mathematics of Physics and Modern Engineering*. McGraw-Hill, New York, 1958.

Sokolnikoff, I. S., and Sokolnikoff, E., *Higher Mathematics for Engineers and Physicists*. McGraw-Hill, New York, 1934.

(B) CALCULUS AND ALGEBRA: TEXTBOOKS IN PURE MATHEMATICS

Birkhoff, G., and MacLane, S., *A Survey of Modern Algebra*, 3rd ed., revised. Macmillan, New York, 1965.

Buck, R. C., *Advanced Calculus*. McGraw-Hill, New York, 1956.

Courant, R., *Differential and Integral Calculus*, Vols. I and II, 2nd English ed. Interscience, New York, 1937.

Thomas, G. B., *Calculus and Analytic Geometry*, 2nd ed. Addison-Wesley, Reading, Massachusetts, 1953.

Weiss, M. J., *Higher Algebra*. Wiley, New York, 1949.

(C) CLASSICAL MECHANICS

Goldstein, H. J., *Classical Mechanics*. Addison-Wesley, Reading, Massachusetts, 1959.

Landau, L. D., and Lifschitz, E. M., *Mechanics*. Translated from Russian by J. B. Sykes and J. S. Bell (Volume 1 in *Course of Theoretical Physics*). Addison-Wesley, Reading, Massachusetts, 1958.

Synge, J. L., and Griffith, B. A., *Principles of Mechanics*, 3rd ed. McGraw-Hill, New York, 1959.

(D) QUANTUM MECHANICS: TEXTS WITH CHAPTERS ON CLASSICAL MECHANICS AND MATHEMATICAL BACKGROUND

Dicke, R. H., and Wittke, J. P., *Introduction to Quantum Mechanics*. Addison Wesley, Reading, Massachusetts, 1960.

Dirac, P. A. M., *The Principles of Quantum Mechanics*. Oxford Univ. Press, New York, 1947.

Eyring, H., Walter, J., and Kimball, G. E., *Quantum Chemistry*. Wiley, New York, 1944.

Kauzmann, W., *Quantum Chemistry*. Academic Press, New York, 1957.

Landau, L. D., and Lifschitz, E. M., *Quantum Mechanics: Nonrelativistic Theory*. Translated from Russian by J. B. Sykes and J. S. Bell (Volume 3 in *Course of Theoretical Physics*). Addison-Wesley, Reading, Massachusetts, 1958.

Pauling, L. C., and Wilson, E. B., *Introduction to Quantum Mechanics*. McGraw-Hill, New York, 1935.

Schiff, L. I., *Quantum Mechanics*. McGraw-Hill, New York, 1955.

Answers to Problems

Chapter 1

1. Solve $df/dx = kf$; solutions are all functions $\{e^{kx}\}$.

Chapter 2

1. Evaluate all integrals of Eq. 2–16.
2. (a) even; (b) odd; (c) even; (d) odd; (e) even; (f) odd; (g) $e^x = \sinh x + \cosh x$, $\sinh x$ is odd, $\cosh x$ is even; (h) $e^{ix} = i \sin x + \cos x$, $\sin x$ is odd, $\cos x$ is even.
3. The norms are $\pi/2$, except $N(\sin 0x) = 0$, $N(\cos 0x) = \pi$. The expansion coefficients are $b_n = (2/\pi)\langle \sin nx \mid f\rangle$, $a_n = (2/\pi)\langle \cos nx \mid f\rangle$ in the series $f = \Sigma b_n \sin nx = a_0/2 + \Sigma a_n \cos nx$.
4. $$\pi x - x^2 = -\sum_{m \text{ odd}} \frac{8}{m^3 \pi} \sin mx$$
$$= \frac{\pi^2}{6} - \sum_{m \text{ even}} \frac{4}{m^2} \cos mx$$

Sine series converges faster.

5. $\sin x = -\sum_{m \text{ odd}} [8/\pi m(m^2 - 4)] \sin mx = \frac{1}{2} - \frac{1}{2} \cos 2x$.

Cosine "series" is the usual trigonometric identity.

6. $\pi^3/12 + \pi/4$.
8. Square wave is odd, so must involve only sine terms:
$$f(x) = \sum_{m \text{ odd}} (4a/m\pi) \sin mx.$$
9. Triangular wave is even, so must involve only cosine terms:
$$f(x) = \pi/2 + \sum_{n \text{ odd}} (4/\pi n^2) \cos nx.$$

11. Taylor series is not an expansion in orthonormal functions. The Schmidt process using different intervals and different orders of the powers of x will give different orthonormal functions.

12. Take the functions in this order: $f_0 = \sin^0 x$, $f_1 = \cos^0 x$, $f_2 = \sin^1 x$, $f_3 = \cos^1 x$, and so forth, and use the Schmidt process.

13. Multiply Eq. 2-114 for $P_l{}^m$ by $P_{l'}{}^m$; multiply the equation for $P_{l'}{}^m$ by $P_l{}^m$, and subtract. Integrate by parts.

Chapter 3

1. (a), (c), (d) are independent; (b) is dependent.
2. Yes.
6.
$$A^{-1} = \begin{pmatrix} \frac{10}{9} & \frac{5}{9} & -\frac{4}{9} \\ -\frac{8}{9} & -\frac{4}{9} & \frac{5}{9} \\ -\frac{15}{9} & -\frac{3}{9} & \frac{6}{9} \end{pmatrix} \quad B^{-1} = \begin{pmatrix} 1 & 0 & 0 \\ -2 & 1 & 0 \\ \frac{2}{3} & -\frac{1}{3} & \frac{1}{3} \end{pmatrix}$$
A and B do not commute.

7. Determinant $= 1$.
8. Use successive applications of expansion in cofactors.
9. (a) $x = -17/54$, $y = -63/54$, $z = -31/54$.
   (b) For example, $w = 7z/5$, $2x = y - 8z/5$.
   (c) $z = 5/7$, $y = 2x + 8/7$.
   (d) $x = 1$, $y = 2$, $z = 3$.
10.
$$\begin{pmatrix} \sin\theta\cos\phi & \sin\theta\sin\phi & \cos\theta \\ \cos\theta\cos\phi & \cos\theta\sin\phi & -\sin\theta \\ -\sin\phi & \cos\phi & 0 \end{pmatrix}$$

11. $(0, -3)$, $(-5, 6)$, $(2, -3)$ $(-1, 0)$.
12. (a) Rank 2; (b) rank 3.
13. (a) $U = \begin{pmatrix} \dfrac{1}{\sqrt{2}} & \dfrac{i}{\sqrt{2}} \\ \dfrac{1}{\sqrt{2}} & -\dfrac{i}{\sqrt{2}} \end{pmatrix}$

   (b) $U'^{-1}A^\phi U' = A^\psi = \begin{pmatrix} e^{-i\alpha} & 0 \\ 0 & e^{i\alpha} \end{pmatrix}$

Appendix 149

14. $\tan 2\theta = 2h/(a - b)$.
16. (e) Show $\text{tr}(S^{-1}AS) = \text{tr } A$, where S is unitary.
21. Eigenvalues: 4, 8, 12.
23. Eigenvalues: 0, 0, 2.

Chapter 4

4. Let (a, θ, ϕ) be the coordinates of the center of mass of the rod on the circle, and let $(l/2, \theta', \phi')$ be the coordinates of the masses about the center of the rod. Then, $T = ma^2(\dot{\theta}^2 + \dot{\phi}^2 \sin^2 \phi) + (ml^2/4)(\dot{\theta}'^2 + \dot{\phi}'^2 \sin^2 \theta')$.

5. In spherical coordinates,

$$-mr(\dot{\theta}^2 + \dot{\phi}^2 \sin^2 \theta) + mg \sin \theta = 0$$
$$mr^2\ddot{\theta} - mr^2\dot{\phi}^2 \sin \theta \cos \theta + mgr \cos \theta = 0$$
$$mr^2\ddot{\phi} \sin^2 \theta + 2mr^2\dot{\theta} \sin \theta \cos \theta = 0$$

6. Use spherical coordinates about the point where the string passes over the table.

7. Using polar coordinates,

$$m\ddot{r} - mr\dot{\theta}^2 + k(r - r_0) = 0$$
$$2m\dot{r}\dot{\theta} + mr^2\ddot{\theta} = 0$$

Angular momentum is conserved.

9. $\omega_1 = (2k_1/m)^{1/2}$, $\omega_2 = (2k_2/m)^{1/2}$. Motion in different directions is not coupled.

15. Hint: Think of orthogonal transformations.

Chapter 5

2. $[x, p_x] = i\hbar$.

Index

Adjoint, 89
Angular momentum, 106
   conservation of, 107, 110
   of a system, 110
Associated Legendre functions, 42, 45

Basis set, 52
   change of, 89
   orthonormal, 52

Center of mass, 109
Characteristic value (*see* Eigenvalue)
Cofactor, 64
Commutator, 89
Completeness, 13
Components of a vector, 48
Congruent transformation, 123
Conjugate coordinates and momenta, 117
Conservation laws, 105–110
   generalized, 115
   with Poisson bracket, 120
   in quantum mechanics, 139
Conservative force (*see* Force, conservative)
Continuum, 2
Coordinate, generalized (*see* Generalized coordinates)
Correspondence principle, 138
Cramer's rule, 63, 70
Cross product, 106

DeBroglie, 137
Degeneracy, 3, 92, 99
Degenerate eigenvalue (*see* Eigenvalue, degenerate)
Degree of degeneracy (*see* Degeneracy)
Degree of freedom, 121
Del, 108
Delta symbol (*see* Kronecker delta symbol)
Determinant, 62–69

evaluation of, 63
expansion of, 64, 65
and linear independence, 69
and matrix inversion, 68
Divergence, 108
Dot product (*see* Scalar product)

Eigenfunction, 2
Eigenvalue, 1
  in classical mechanics, 4, 123
  degenerate, 3
  in quantum mechanics, 2
Eigenvalue equation, matrix form, 95
Eigenvector, 4
Element of a matrix (*see* Matrix element)
Elementary row operations, 55
  effect on determinant, 67
  and linear independence, 55
  and matrix inversion, 61
Energy, conservation of, 109, 110
  and Hamiltonian, 119
  (*see also* Kinetic energy, Potential)
Euclidean vector space (*see* Vector space, Euclidean)
Expansion, in Fourier series, 22
  of even and odd functions, 22
  in orthonormal functions, 15–21, 22
  error incurred in, 17
  of a vector, 53, 99
Expansion interval, 7
Expansion theorem, 20, 53

Force, conservative, 108
  generalized, 112
Fourier series, 21–27
  expansion in terms of (*see* Expansion)
  in exponentials, 26 (*see also* Harmonic functions)
  "half Fourier series," 25
  on symmetric interval, 25
Function, 7
  associated Legendre (*see* Associated Legendre function)
  even, 10
  integral of, 10
  harmonic (*see* Harmonic functions)
  Hermite (*see* Hermite functions)
  Laguerre (*see* Laguerre functions)
  normalizable, 30
  odd, 10
  integral of, 10
  orthogonal (*see* Orthogonal functions)
  set of (*see* Sets of functions)
  square-integrable (*see* Function, normalizable)

Generalized coordinates, 110–121
Generalized force (*see* Force, generalized)
Generalized momentum, 117
Generating function, 35
Gradient, 108

Hamiltonian, 117
  conservation of, 118
  and total energy, 119
Hamilton's equations, 118, 120
Harmonic functions, 44
Harmonic oscillator, 121
Heisenberg equation, 139
Heisenberg matrix mechanics (*see* Matrix mechanics)
Hermite functions, 45
Hermite polynomials, 32
Hermitian vector space (*see* Vector space, Hermitian)
Hilbert space, 141

Index

Hooke's law, 121

Index of set of functions, 13
Inertia (*see* Moment of inertia)
Inner product, bilinear, 49
  of even and odd functions, 13
  of functions, 7
  Hermitian, 49
  positive, 49
  symmetric, 49
  of vectors, 49, 106
Interval (*see* Expansion interval)

Kinetic energy, 108
  of rotation, 131
  of a system, 110
Kronecker delta symbol, 14, 17

Lagrange's equations, 113, 115, 120
Lagrangian, 115
Laguerre functions, 45
Laplacian, 108
Legendre differential equation, 33
Legendre polynomials, 31–45
  evaluation at boundary points, 38
  generating function for, 35
  recursion relations among, 40
  Rodrigues' formula for, 36
  by Schmidt orthogonalization, 32
  solution to Legendre differential equation, 33
  (*see also* Associated Legendre functions, Spherical harmonics)
Linear equations, 69–76
  homogeneous, 70
  matrix form, 69
Linear independence, 27, 52, 53, 69
  of vectors, test for, 55, 69
Linear momentum (*see* Momentum)
Linear operator (*see* Operator, linear)

Linear transformation, 76–84

Map, 78
Matrix, 3, 55, 57–62
  augmented, 70
  coefficient, 62
  diagonal, 55, 96
  identity, 60
  inverse, 60
  null, 59
  product of, 57
  sum of, 57
  transpose, 62
  triangular, 55
  unit, 60
  zero, 59
Matrix element, 55, 87
Matrix equation, 69
Matrix mechanics, 3
Minor, 64
Moment of inertia, 131
  principal, 132
Momentum, 106
  conservation of, 107, 110
  generalized (*see* Generalized momentum)
  of a system, 110
  (*see also* Angular momentum)

Nabla, 108
Newton's laws, 107
Norm, 8, 50
Normalization, 9, 50

Operator, 2, 86, 99
  Hermitian, 92, 98
  linear, 84–101
  matrix representation of, 88
  orthogonal, 94
  projection, 90, 99
  symmetric, 94

unitary, 92
Order of spherical harmonics, 43
Orthogonal functions, 6–47
Orthogonal matrix, 81
Orthogonal transformation, 79, 89
Orthogonality, 8, 50
Orthonormal functions, construction of, 27–32
  expansion in terms of (*see* Expansion)
Orthonormality, 14, 50

Permutation, 63
  sign of, 63
Poisson bracket, 120
Potential, 108
Power series (*see* Expansion)
Principal axis transformation, 132
Principal moment of inertia, 132
Projection operator, 90, 99

Rank, of Legendre polynomials, 33
  of linear transformation, 77
  of matrix, 77
  of spherical harmonics, 43
Recursion relation, for Legendre polynomials, 35
  for power series coefficients, 35
Rodrigues' formula (*see* Legendre polynomials)
Rotation, 81–84, 129–134
Row equivalence, 55

Sawtooth wave, 24
  Fourier synthesis of, 26
Scalar product, 8, 106
  (*see also* Inner product)
Schmidt orthogonalization, 28, 52
Schrödinger equation, 137, 140
Schrödinger wave mechanics (*see* Wave mechanics)

Schwarz's inequality, 50
Secular equation, 94, 121
Selection rules, 46
Separation theorems, 109–110
Sets of functions, 13
  complete (*see* Completeness)
  orthonormal, 14
Similarity transformation, 89, 94, 96
Spherical harmonics, 42
Square wave, 46

Torque, 107
Trace, 90
Transformation, 77
  (*see also* Congruent transformation, Linear transformation, Orthogonal transformation, Similarity transformation, Unitary transformation)
Transition dipole moment, 47
Triangular wave, 46

Unitary matrix, 81
Unitary transformation, 79, 89

Vector, 3, 48
  matrix representation of, 88
Vector product with scalar, 49
Vector space, 49
  basis set for (*see* Basis set)
  dimension of, 52
  Euclidean, 49, 105
  Hermitian, 49
  spanned, 51
Vector sum, 49
Vibration, 121–129

Wave function, 141
Wave mechanics, 3
Work, 108

A CATALOG OF SELECTED
DOVER BOOKS
IN SCIENCE AND MATHEMATICS

CATALOG OF DOVER BOOKS

Astronomy

BURNHAM'S CELESTIAL HANDBOOK, Robert Burnham, Jr. Thorough guide to the stars beyond our solar system. Exhaustive treatment. Alphabetical by constellation: Andromeda to Cetus in Vol. 1; Chamaeleon to Orion in Vol. 2; and Pavo to Vulpecula in Vol. 3. Hundreds of illustrations. Index in Vol. 3. 2,000pp. 6⅛ x 9¼.
Vol. I: 23567-X
Vol. II: 23568-8
Vol. III: 23673-0

EXPLORING THE MOON THROUGH BINOCULARS AND SMALL TELESCOPES, Ernest H. Cherrington, Jr. Informative, profusely illustrated guide to locating and identifying craters, rills, seas, mountains, other lunar features. Newly revised and updated with special section of new photos. Over 100 photos and diagrams. 240pp. 8¼ x 11. 24491-1

THE EXTRATERRESTRIAL LIFE DEBATE, 1750–1900, Michael J. Crowe. First detailed, scholarly study in English of the many ideas that developed from 1750 to 1900 regarding the existence of intelligent extraterrestrial life. Examines ideas of Kant, Herschel, Voltaire, Percival Lowell, many other scientists and thinkers. 16 illustrations. 704pp. 5⅜ x 8½. 40675-X

THEORIES OF THE WORLD FROM ANTIQUITY TO THE COPERNICAN REVOLUTION, Michael J. Crowe. Newly revised edition of an accessible, enlightening book recreates the change from an earth-centered to a sun-centered conception of the solar system. 242pp. 5⅜ x 8½. 41444-2

A HISTORY OF ASTRONOMY, A. Pannekoek. Well-balanced, carefully reasoned study covers such topics as Ptolemaic theory, work of Copernicus, Kepler, Newton, Eddington's work on stars, much more. Illustrated. References. 521pp. 5⅜ x 8½.
65994-1

A COMPLETE MANUAL OF AMATEUR ASTRONOMY: Tools and Techniques for Astronomical Observations, P. Clay Sherrod with Thomas L. Koed. Concise, highly readable book discusses: selecting, setting up and maintaining a telescope; amateur studies of the sun; lunar topography and occultations; observations of Mars, Jupiter, Saturn, the minor planets and the stars; an introduction to photoelectric photometry; more. 1981 ed. 124 figures. 26 halftones. 37 tables. 335pp. 6½ x 9¼.
42820-6

AMATEUR ASTRONOMER'S HANDBOOK, J. B. Sidgwick. Timeless, comprehensive coverage of telescopes, mirrors, lenses, mountings, telescope drives, micrometers, spectroscopes, more. 189 illustrations. 576pp. 5⅜ x 8¼. (Available in U.S. only.)
24034-7

STARS AND RELATIVITY, Ya. B. Zel'dovich and I. D. Novikov. Vol. 1 of *Relativistic Astrophysics* by famed Russian scientists. General relativity, properties of matter under astrophysical conditions, stars, and stellar systems. Deep physical insights, clear presentation. 1971 edition. References. 544pp. 5⅜ x 8¼. 69424-0

CATALOG OF DOVER BOOKS

Chemistry

THE SCEPTICAL CHYMIST: The Classic 1661 Text, Robert Boyle. Boyle defines the term "element," asserting that all natural phenomena can be explained by the motion and organization of primary particles. 1911 ed. viii+232pp. 5⅜ x 8½.
42825-7

RADIOACTIVE SUBSTANCES, Marie Curie. Here is the celebrated scientist's doctoral thesis, the prelude to her receipt of the 1903 Nobel Prize. Curie discusses establishing atomic character of radioactivity found in compounds of uranium and thorium; extraction from pitchblende of polonium and radium; isolation of pure radium chloride; determination of atomic weight of radium; plus electric, photographic, luminous, heat, color effects of radioactivity. ii+94pp. 5⅜ x 8½. 42550-9

CHEMICAL MAGIC, Leonard A. Ford. Second Edition, Revised by E. Winston Grundmeier. Over 100 unusual stunts demonstrating cold fire, dust explosions, much more. Text explains scientific principles and stresses safety precautions. 128pp. 5⅜ x 8½.
67628-5

THE DEVELOPMENT OF MODERN CHEMISTRY, Aaron J. Ihde. Authoritative history of chemistry from ancient Greek theory to 20th-century innovation. Covers major chemists and their discoveries. 209 illustrations. 14 tables. Bibliographies. Indices. Appendices. 851pp. 5⅜ x 8½. 64235-6

CATALYSIS IN CHEMISTRY AND ENZYMOLOGY, William P. Jencks. Exceptionally clear coverage of mechanisms for catalysis, forces in aqueous solution, carbonyl- and acyl-group reactions, practical kinetics, more. 864pp. 5⅜ x 8½.
65460-5

ELEMENTS OF CHEMISTRY, Antoine Lavoisier. Monumental classic by founder of modern chemistry in remarkable reprint of rare 1790 Kerr translation. A must for every student of chemistry or the history of science. 539pp. 5⅜ x 8½. 64624-6

THE HISTORICAL BACKGROUND OF CHEMISTRY, Henry M. Leicester. Evolution of ideas, not individual biography. Concentrates on formulation of a coherent set of chemical laws. 260pp. 5⅜ x 8½. 61053-5

A SHORT HISTORY OF CHEMISTRY, J. R. Partington. Classic exposition explores origins of chemistry, alchemy, early medical chemistry, nature of atmosphere, theory of valency, laws and structure of atomic theory, much more. 428pp. 5⅜ x 8½. (Available in U.S. only.) 65977-1

GENERAL CHEMISTRY, Linus Pauling. Revised 3rd edition of classic first-year text by Nobel laureate. Atomic and molecular structure, quantum mechanics, statistical mechanics, thermodynamics correlated with descriptive chemistry. Problems. 992pp. 5⅜ x 8½. 65622-5

FROM ALCHEMY TO CHEMISTRY, John Read. Broad, humanistic treatment focuses on great figures of chemistry and ideas that revolutionized the science. 50 illustrations. 240pp. 5⅜ x 8½. 28690-8

CATALOG OF DOVER BOOKS

Engineering

DE RE METALLICA, Georgius Agricola. The famous Hoover translation of greatest treatise on technological chemistry, engineering, geology, mining of early modern times (1556). All 289 original woodcuts. 638pp. 6¾ x 11. 60006-8

FUNDAMENTALS OF ASTRODYNAMICS, Roger Bate et al. Modern approach developed by U.S. Air Force Academy. Designed as a first course. Problems, exercises. Numerous illustrations. 455pp. 5⅜ x 8½. 60061-0

DYNAMICS OF FLUIDS IN POROUS MEDIA, Jacob Bear. For advanced students of ground water hydrology, soil mechanics and physics, drainage and irrigation engineering, and more. 335 illustrations. Exercises, with answers. 784pp. 6⅛ x 9¼. 65675-6

THEORY OF VISCOELASTICITY (Second Edition), Richard M. Christensen. Complete, consistent description of the linear theory of the viscoelastic behavior of materials. Problem-solving techniques discussed. 1982 edition. 29 figures. xiv+364pp. 6⅛ x 9¼. 42880-X

MECHANICS, J. P. Den Hartog. A classic introductory text or refresher. Hundreds of applications and design problems illuminate fundamentals of trusses, loaded beams and cables, etc. 334 answered problems. 462pp. 5⅜ x 8½. 60754-2

MECHANICAL VIBRATIONS, J. P. Den Hartog. Classic textbook offers lucid explanations and illustrative models, applying theories of vibrations to a variety of practical industrial engineering problems. Numerous figures. 233 problems, solutions. Appendix. Index. Preface. 436pp. 5⅜ x 8½. 64785-4

STRENGTH OF MATERIALS, J. P. Den Hartog. Full, clear treatment of basic material (tension, torsion, bending, etc.) plus advanced material on engineering methods, applications. 350 answered problems. 323pp. 5⅜ x 8½. 60755-0

A HISTORY OF MECHANICS, René Dugas. Monumental study of mechanical principles from antiquity to quantum mechanics. Contributions of ancient Greeks, Galileo, Leonardo, Kepler, Lagrange, many others. 671pp. 5⅜ x 8½. 65632-2

STABILITY THEORY AND ITS APPLICATIONS TO STRUCTURAL MECHANICS, Clive L. Dym. Self-contained text focuses on Koiter postbuckling analyses, with mathematical notions of stability of motion. Basing minimum energy principles for static stability upon dynamic concepts of stability of motion, it develops asymptotic buckling and postbuckling analyses from potential energy considerations, with applications to columns, plates, and arches. 1974 ed. 208pp. 5⅜ x 8½. 42541-X

METAL FATIGUE, N. E. Frost, K. J. Marsh, and L. P. Pook. Definitive, clearly written, and well-illustrated volume addresses all aspects of the subject, from the historical development of understanding metal fatigue to vital concepts of the cyclic stress that causes a crack to grow. Includes 7 appendixes. 544pp. 5⅜ x 8½. 40927-9

CATALOG OF DOVER BOOKS

ROCKETS, Robert Goddard. Two of the most significant publications in the history of rocketry and jet propulsion: "A Method of Reaching Extreme Altitudes" (1919) and "Liquid Propellant Rocket Development" (1936). 128pp. 5⅜ x 8½. 42537-1

STATISTICAL MECHANICS: Principles and Applications, Terrell L. Hill. Standard text covers fundamentals of statistical mechanics, applications to fluctuation theory, imperfect gases, distribution functions, more. 448pp. 5⅜ x 8½. 65390-0

ENGINEERING AND TECHNOLOGY 1650–1750: Illustrations and Texts from Original Sources, Martin Jensen. Highly readable text with more than 200 contemporary drawings and detailed engravings of engineering projects dealing with surveying, leveling, materials, hand tools, lifting equipment, transport and erection, piling, bailing, water supply, hydraulic engineering, and more. Among the specific projects outlined–transporting a 50-ton stone to the Louvre, erecting an obelisk, building timber locks, and dredging canals. 207pp. 8⅜ x 11¼. 42232-1

THE VARIATIONAL PRINCIPLES OF MECHANICS, Cornelius Lanczos. Graduate level coverage of calculus of variations, equations of motion, relativistic mechanics, more. First inexpensive paperbound edition of classic treatise. Index. Bibliography. 418pp. 5⅜ x 8½. 65067-7

PROTECTION OF ELECTRONIC CIRCUITS FROM OVERVOLTAGES, Ronald B. Standler. Five-part treatment presents practical rules and strategies for circuits designed to protect electronic systems from damage by transient overvoltages. 1989 ed. xxiv+434pp. 6⅛ x 9¼. 42552-5

ROTARY WING AERODYNAMICS, W. Z. Stepniewski. Clear, concise text covers aerodynamic phenomena of the rotor and offers guidelines for helicopter performance evaluation. Originally prepared for NASA. 537 figures. 640pp. 6⅛ x 9¼. 64647-5

INTRODUCTION TO SPACE DYNAMICS, William Tyrrell Thomson. Comprehensive, classic introduction to space-flight engineering for advanced undergraduate and graduate students. Includes vector algebra, kinematics, transformation of coordinates. Bibliography. Index. 352pp. 5⅜ x 8½. 65113-4

HISTORY OF STRENGTH OF MATERIALS, Stephen P. Timoshenko. Excellent historical survey of the strength of materials with many references to the theories of elasticity and structure. 245 figures. 452pp. 5⅜ x 8½. 61187-6

ANALYTICAL FRACTURE MECHANICS, David J. Unger. Self-contained text supplements standard fracture mechanics texts by focusing on analytical methods for determining crack-tip stress and strain fields. 336pp. 6⅛ x 9¼. 41737-9

STATISTICAL MECHANICS OF ELASTICITY, J. H. Weiner. Advanced, self-contained treatment illustrates general principles and elastic behavior of solids. Part 1, based on classical mechanics, studies thermoelastic behavior of crystalline and polymeric solids. Part 2, based on quantum mechanics, focuses on interatomic force laws, behavior of solids, and thermally activated processes. For students of physics and chemistry and for polymer physicists. 1983 ed. 96 figures. 496pp. 5⅜ x 8½. 42260-7

Mathematics

FUNCTIONAL ANALYSIS (Second Corrected Edition), George Bachman and Lawrence Narici. Excellent treatment of subject geared toward students with background in linear algebra, advanced calculus, physics, and engineering. Text covers introduction to inner-product spaces, normed, metric spaces, and topological spaces; complete orthonormal sets, the Hahn-Banach Theorem and its consequences, and many other related subjects. 1966 ed. 544pp. 6⅛ x 9¼. 40251-7

ASYMPTOTIC EXPANSIONS OF INTEGRALS, Norman Bleistein & Richard A. Handelsman. Best introduction to important field with applications in a variety of scientific disciplines. New preface. Problems. Diagrams. Tables. Bibliography. Index. 448pp. 5⅜ x 8½. 65082-0

VECTOR AND TENSOR ANALYSIS WITH APPLICATIONS, A. I. Borisenko and I. E. Tarapov. Concise introduction. Worked-out problems, solutions, exercises. 257pp. 5⅝ x 8¼. 63833-2

THE ABSOLUTE DIFFERENTIAL CALCULUS (CALCULUS OF TENSORS), Tullio Levi-Civita. Great 20th-century mathematician's classic work on material necessary for mathematical grasp of theory of relativity. 452pp. 5⅝ x 8¼. 63401-9

AN INTRODUCTION TO ORDINARY DIFFERENTIAL EQUATIONS, Earl A. Coddington. A thorough and systematic first course in elementary differential equations for undergraduates in mathematics and science, with many exercises and problems (with answers). Index. 304pp. 5⅜ x 8½. 65942-9

FOURIER SERIES AND ORTHOGONAL FUNCTIONS, Harry F. Davis. An incisive text combining theory and practical example to introduce Fourier series, orthogonal functions and applications of the Fourier method to boundary-value problems. 570 exercises. Answers and notes. 416pp. 5⅜ x 8½. 65973-9

COMPUTABILITY AND UNSOLVABILITY, Martin Davis. Classic graduate-level introduction to theory of computability, usually referred to as theory of recurrent functions. New preface and appendix. 288pp. 5⅜ x 8½. 61471-9

ASYMPTOTIC METHODS IN ANALYSIS, N. G. de Bruijn. An inexpensive, comprehensive guide to asymptotic methods–the pioneering work that teaches by explaining worked examples in detail. Index. 224pp. 5⅜ x 8½ 64221-6

APPLIED COMPLEX VARIABLES, John W. Dettman. Step-by-step coverage of fundamentals of analytic function theory–plus lucid exposition of five important applications: Potential Theory; Ordinary Differential Equations; Fourier Transforms; Laplace Transforms; Asymptotic Expansions. 66 figures. Exercises at chapter ends. 512pp. 5⅜ x 8½. 64670-X

INTRODUCTION TO LINEAR ALGEBRA AND DIFFERENTIAL EQUATIONS, John W. Dettman. Excellent text covers complex numbers, determinants, orthonormal bases, Laplace transforms, much more. Exercises with solutions. Undergraduate level. 416pp. 5⅜ x 8½. 65191-6

CATALOG OF DOVER BOOKS

CALCULUS OF VARIATIONS WITH APPLICATIONS, George M. Ewing. Applications-oriented introduction to variational theory develops insight and promotes understanding of specialized books, research papers. Suitable for advanced undergraduate/graduate students as primary, supplementary text. 352pp. 5⅜ x 8½.
64856-7

COMPLEX VARIABLES, Francis J. Flanigan. Unusual approach, delaying complex algebra till harmonic functions have been analyzed from real variable viewpoint. Includes problems with answers. 364pp. 5⅜ x 8½. 61388-7

AN INTRODUCTION TO THE CALCULUS OF VARIATIONS, Charles Fox. Graduate-level text covers variations of an integral, isoperimetrical problems, least action, special relativity, approximations, more. References. 279pp. 5⅜ x 8½.
65499-0

COUNTEREXAMPLES IN ANALYSIS, Bernard R. Gelbaum and John M. H. Olmsted. These counterexamples deal mostly with the part of analysis known as "real variables." The first half covers the real number system, and the second half encompasses higher dimensions. 1962 edition. xxiv+198pp. 5⅜ x 8½. 42875-3

CATASTROPHE THEORY FOR SCIENTISTS AND ENGINEERS, Robert Gilmore. Advanced-level treatment describes mathematics of theory grounded in the work of Poincaré, R. Thom, other mathematicians. Also important applications to problems in mathematics, physics, chemistry, and engineering. 1981 edition. References. 28 tables. 397 black-and-white illustrations. xvii+666pp. 6⅛ x 9¼.
67539-4

INTRODUCTION TO DIFFERENCE EQUATIONS, Samuel Goldberg. Exceptionally clear exposition of important discipline with applications to sociology, psychology, economics. Many illustrative examples; over 250 problems. 260pp. 5⅜ x 8½.
65084-7

NUMERICAL METHODS FOR SCIENTISTS AND ENGINEERS, Richard Hamming. Classic text stresses frequency approach in coverage of algorithms, polynomial approximation, Fourier approximation, exponential approximation, other topics. Revised and enlarged 2nd edition. 721pp. 5⅜ x 8½. 65241-6

INTRODUCTION TO NUMERICAL ANALYSIS (2nd Edition), F. B. Hildebrand. Classic, fundamental treatment covers computation, approximation, interpolation, numerical differentiation and integration, other topics. 150 new problems. 669pp. 5⅜ x 8½. 65363-3

THREE PEARLS OF NUMBER THEORY, A. Y. Khinchin. Three compelling puzzles require proof of a basic law governing the world of numbers. Challenges concern van der Waerden's theorem, the Landau-Schnirelmann hypothesis and Mann's theorem, and a solution to Waring's problem. Solutions included. 64pp. 5⅜ x 8½.
40026-3

THE PHILOSOPHY OF MATHEMATICS: An Introductory Essay, Stephan Körner. Surveys the views of Plato, Aristotle, Leibniz & Kant concerning propositions and theories of applied and pure mathematics. Introduction. Two appendices. Index. 198pp. 5⅜ x 8½. 25048-2

CATALOG OF DOVER BOOKS

INTRODUCTORY REAL ANALYSIS, A.N. Kolmogorov, S. V. Fomin. Translated by Richard A. Silverman. Self-contained, evenly paced introduction to real and functional analysis. Some 350 problems. 403pp. 5⅜ x 8½. 61226-0

APPLIED ANALYSIS, Cornelius Lanczos. Classic work on analysis and design of finite processes for approximating solution of analytical problems. Algebraic equations, matrices, harmonic analysis, quadrature methods, more. 559pp. 5⅜ x 8½. 65656-X

AN INTRODUCTION TO ALGEBRAIC STRUCTURES, Joseph Landin. Superb self-contained text covers "abstract algebra": sets and numbers, theory of groups, theory of rings, much more. Numerous well-chosen examples, exercises. 247pp. 5⅜ x 8½. 65940-2

QUALITATIVE THEORY OF DIFFERENTIAL EQUATIONS, V. V. Nemytskii and V.V. Stepanov. Classic graduate-level text by two prominent Soviet mathematicians covers classical differential equations as well as topological dynamics and ergodic theory. Bibliographies. 523pp. 5⅜ x 8½. 65954-2

THEORY OF MATRICES, Sam Perlis. Outstanding text covering rank, nonsingularity and inverses in connection with the development of canonical matrices under the relation of equivalence, and without the intervention of determinants. Includes exercises. 237pp. 5⅜ x 8½. 66810-X

INTRODUCTION TO ANALYSIS, Maxwell Rosenlicht. Unusually clear, accessible coverage of set theory, real number system, metric spaces, continuous functions, Riemann integration, multiple integrals, more. Wide range of problems. Undergraduate level. Bibliography. 254pp. 5⅜ x 8½. 65038-3

MODERN NONLINEAR EQUATIONS, Thomas L. Saaty. Emphasizes practical solution of problems; covers seven types of equations. ". . . a welcome contribution to the existing literature. . . . "–*Math Reviews*. 490pp. 5⅜ x 8½. 64232-1

MATRICES AND LINEAR ALGEBRA, Hans Schneider and George Phillip Barker. Basic textbook covers theory of matrices and its applications to systems of linear equations and related topics such as determinants, eigenvalues, and differential equations. Numerous exercises. 432pp. 5⅜ x 8½. 66014-1

MATHEMATICS APPLIED TO CONTINUUM MECHANICS, Lee A. Segel. Analyzes models of fluid flow and solid deformation. For upper-level math, science, and engineering students. 608pp. 5⅜ x 8½. 65369-2

ELEMENTS OF REAL ANALYSIS, David A. Sprecher. Classic text covers fundamental concepts, real number system, point sets, functions of a real variable, Fourier series, much more. Over 500 exercises. 352pp. 5⅜ x 8½. 65385-4

SET THEORY AND LOGIC, Robert R. Stoll. Lucid introduction to unified theory of mathematical concepts. Set theory and logic seen as tools for conceptual understanding of real number system. 496pp. 5⅜ x 8¼. 63829-4

CATALOG OF DOVER BOOKS

TENSOR CALCULUS, J.L. Synge and A. Schild. Widely used introductory text covers spaces and tensors, basic operations in Riemannian space, non-Riemannian spaces, etc. 324pp. 5⅜ x 8¼. 63612-7

ORDINARY DIFFERENTIAL EQUATIONS, Morris Tenenbaum and Harry Pollard. Exhaustive survey of ordinary differential equations for undergraduates in mathematics, engineering, science. Thorough analysis of theorems. Diagrams. Bibliography. Index. 818pp. 5⅜ x 8½. 64940-7

INTEGRAL EQUATIONS, F. G. Tricomi. Authoritative, well-written treatment of extremely useful mathematical tool with wide applications. Volterra Equations, Fredholm Equations, much more. Advanced undergraduate to graduate level. Exercises. Bibliography. 238pp. 5⅜ x 8½. 64828-1

FOURIER SERIES, Georgi P. Tolstov. Translated by Richard A. Silverman. A valuable addition to the literature on the subject, moving clearly from subject to subject and theorem to theorem. 107 problems, answers. 336pp. 5⅜ x 8½. 63317-9

INTRODUCTION TO MATHEMATICAL THINKING, Friedrich Waismann. Examinations of arithmetic, geometry, and theory of integers; rational and natural numbers; complete induction; limit and point of accumulation; remarkable curves; complex and hypercomplex numbers, more. 1959 ed. 27 figures. xii+260pp. 5⅜ x 8½. 42804-4

POPULAR LECTURES ON MATHEMATICAL LOGIC, Hao Wang. Noted logician's lucid treatment of historical developments, set theory, model theory, recursion theory and constructivism, proof theory, more. 3 appendixes. Bibliography. 1981 ed. ix+283pp. 5⅜ x 8½. 67632-3

CALCULUS OF VARIATIONS, Robert Weinstock. Basic introduction covering isoperimetric problems, theory of elasticity, quantum mechanics, electrostatics, etc. Exercises throughout. 326pp. 5⅜ x 8½. 63069-2

THE CONTINUUM: A Critical Examination of the Foundation of Analysis, Hermann Weyl. Classic of 20th-century foundational research deals with the conceptual problem posed by the continuum. 156pp. 5⅜ x 8½. 67982-9

CHALLENGING MATHEMATICAL PROBLEMS WITH ELEMENTARY SOLUTIONS, A. M. Yaglom and I. M. Yaglom. Over 170 challenging problems on probability theory, combinatorial analysis, points and lines, topology, convex polygons, many other topics. Solutions. Total of 445pp. 5⅜ x 8½. Two-vol. set.
Vol. I: 65536-9   Vol. II: 65537-7

INTRODUCTION TO PARTIAL DIFFERENTIAL EQUATIONS WITH APPLICATIONS, E. C. Zachmanoglou and Dale W. Thoe. Essentials of partial differential equations applied to common problems in engineering and the physical sciences. Problems and answers. 416pp. 5⅜ x 8½. 65251-3

THE THEORY OF GROUPS, Hans J. Zassenhaus. Well-written graduate-level text acquaints reader with group-theoretic methods and demonstrates their usefulness in mathematics. Axioms, the calculus of complexes, homomorphic mapping, p-group theory, more. 276pp. 5⅜ x 8½. 40922-8

CATALOG OF DOVER BOOKS

Math–Decision Theory, Statistics, Probability

ELEMENTARY DECISION THEORY, Herman Chernoff and Lincoln E. Moses. Clear introduction to statistics and statistical theory covers data processing, probability and random variables, testing hypotheses, much more. Exercises. 364pp. 5⅜ x 8½. 65218-1

STATISTICS MANUAL, Edwin L. Crow et al. Comprehensive, practical collection of classical and modern methods prepared by U.S. Naval Ordnance Test Station. Stress on use. Basics of statistics assumed. 288pp. 5⅜ x 8½. 60599-X

SOME THEORY OF SAMPLING, William Edwards Deming. Analysis of the problems, theory, and design of sampling techniques for social scientists, industrial managers, and others who find statistics important at work. 61 tables. 90 figures. xvii +602pp. 5⅜ x 8½. 64684-X

LINEAR PROGRAMMING AND ECONOMIC ANALYSIS, Robert Dorfman, Paul A. Samuelson and Robert M. Solow. First comprehensive treatment of linear programming in standard economic analysis. Game theory, modern welfare economics, Leontief input-output, more. 525pp. 5⅜ x 8½. 65491-5

PROBABILITY: An Introduction, Samuel Goldberg. Excellent basic text covers set theory, probability theory for finite sample spaces, binomial theorem, much more. 360 problems. Bibliographies. 322pp. 5⅜ x 8½. 65252-1

GAMES AND DECISIONS: Introduction and Critical Survey, R. Duncan Luce and Howard Raiffa. Superb nontechnical introduction to game theory, primarily applied to social sciences. Utility theory, zero-sum games, n-person games, decision-making, much more. Bibliography. 509pp. 5⅜ x 8½. 65943-7

INTRODUCTION TO THE THEORY OF GAMES, J. C. C. McKinsey. This comprehensive overview of the mathematical theory of games illustrates applications to situations involving conflicts of interest, including economic, social, political, and military contexts. Appropriate for advanced undergraduate and graduate courses; advanced calculus a prerequisite. 1952 ed. x+372pp. 5⅜ x 8½. 42811-7

FIFTY CHALLENGING PROBLEMS IN PROBABILITY WITH SOLUTIONS, Frederick Mosteller. Remarkable puzzlers, graded in difficulty, illustrate elementary and advanced aspects of probability. Detailed solutions. 88pp. 5⅜ x 8½. 65355-2

PROBABILITY THEORY: A Concise Course, Y. A. Rozanov. Highly readable, self-contained introduction covers combination of events, dependent events, Bernoulli trials, etc. 148pp. 5⅜ x 8¼. 63544-9

STATISTICAL METHOD FROM THE VIEWPOINT OF QUALITY CONTROL, Walter A. Shewhart. Important text explains regulation of variables, uses of statistical control to achieve quality control in industry, agriculture, other areas. 192pp. 5⅜ x 8½. 65232-7

Math–Geometry and Topology

ELEMENTARY CONCEPTS OF TOPOLOGY, Paul Alexandroff. Elegant, intuitive approach to topology from set-theoretic topology to Betti groups; how concepts of topology are useful in math and physics. 25 figures. 57pp. 5⅜ x 8½. 60747-X

COMBINATORIAL TOPOLOGY, P. S. Alexandrov. Clearly written, well-organized, three-part text begins by dealing with certain classic problems without using the formal techniques of homology theory and advances to the central concept, the Betti groups. Numerous detailed examples. 654pp. 5⅜ x 8½. 40179-0

EXPERIMENTS IN TOPOLOGY, Stephen Barr. Classic, lively explanation of one of the byways of mathematics. Klein bottles, Moebius strips, projective planes, map coloring, problem of the Koenigsberg bridges, much more, described with clarity and wit. 43 figures. 210pp. 5⅜ x 8½. 25933-1

CONFORMAL MAPPING ON RIEMANN SURFACES, Harvey Cohn. Lucid, insightful book presents ideal coverage of subject. 334 exercises make book perfect for self-study. 55 figures. 352pp. 5⅝ x 8¼. 64025-6

THE GEOMETRY OF RENÉ DESCARTES, René Descartes. The great work founded analytical geometry. Original French text, Descartes's own diagrams, together with definitive Smith-Latham translation. 244pp. 5⅜ x 8½. 60068-8

PRACTICAL CONIC SECTIONS: The Geometric Properties of Ellipses, Parabolas and Hyperbolas, J. W. Downs. This text shows how to create ellipses, parabolas, and hyperbolas. It also presents historical background on their ancient origins and describes the reflective properties and roles of curves in design applications. 1993 ed. 98 figures. xii+100pp. 6½ x 9¼. 42876-1

THE THIRTEEN BOOKS OF EUCLID'S ELEMENTS, translated with introduction and commentary by Thomas L. Heath. Definitive edition. Textual and linguistic notes, mathematical analysis. 2,500 years of critical commentary. Unabridged. 1,414pp. 5⅜ x 8½. Three-vol. set. Vol. I: 60088-2 Vol. II: 60089-0 Vol. III: 60090-4

GEOMETRY OF COMPLEX NUMBERS, Hans Schwerdtfeger. Illuminating, widely praised book on analytic geometry of circles, the Moebius transformation, and two-dimensional non-Euclidean geometries. 200pp. 5⅝ x 8¼. 63830-8

DIFFERENTIAL GEOMETRY, Heinrich W. Guggenheimer. Local differential geometry as an application of advanced calculus and linear algebra. Curvature, transformation groups, surfaces, more. Exercises. 62 figures. 378pp. 5⅜ x 8½. 63433-7

CURVATURE AND HOMOLOGY: Enlarged Edition, Samuel I. Goldberg. Revised edition examines topology of differentiable manifolds; curvature, homology of Riemannian manifolds; compact Lie groups; complex manifolds; curvature, homology of Kaehler manifolds. New Preface. Four new appendixes. 416pp. 5⅜ x 8½. 40207-X

CATALOG OF DOVER BOOKS

History of Math

THE WORKS OF ARCHIMEDES, Archimedes (T. L. Heath, ed.). Topics include the famous problems of the ratio of the areas of a cylinder and an inscribed sphere; the measurement of a circle; the properties of conoids, spheroids, and spirals; and the quadrature of the parabola. Informative introduction. clxxxvi+326pp; supplement, 52pp. 5⅜ x 8½. 42084-1

A SHORT ACCOUNT OF THE HISTORY OF MATHEMATICS, W. W. Rouse Ball. One of clearest, most authoritative surveys from the Egyptians and Phoenicians through 19th-century figures such as Grassman, Galois, Riemann. Fourth edition. 522pp. 5⅜ x 8½. 20630-0

THE HISTORY OF THE CALCULUS AND ITS CONCEPTUAL DEVELOPMENT, Carl B. Boyer. Origins in antiquity, medieval contributions, work of Newton, Leibniz, rigorous formulation. Treatment is verbal. 346pp. 5⅜ x 8½. 60509-4

THE HISTORICAL ROOTS OF ELEMENTARY MATHEMATICS, Lucas N. H. Bunt, Phillip S. Jones, and Jack D. Bedient. Fundamental underpinnings of modern arithmetic, algebra, geometry, and number systems derived from ancient civilizations. 320pp. 5⅜ x 8½. 25563-8

A HISTORY OF MATHEMATICAL NOTATIONS, Florian Cajori. This classic study notes the first appearance of a mathematical symbol and its origin, the competition it encountered, its spread among writers in different countries, its rise to popularity, its eventual decline or ultimate survival. Original 1929 two-volume edition presented here in one volume. xxviii+820pp. 5⅜ x 8½. 67766-4

GAMES, GODS & GAMBLING: A History of Probability and Statistical Ideas, F. N. David. Episodes from the lives of Galileo, Fermat, Pascal, and others illustrate this fascinating account of the roots of mathematics. Features thought-provoking references to classics, archaeology, biography, poetry. 1962 edition. 304pp. 5⅜ x 8½. (Available in U.S. only.) 40023-9

OF MEN AND NUMBERS: The Story of the Great Mathematicians, Jane Muir. Fascinating accounts of the lives and accomplishments of history's greatest mathematical minds–Pythagoras, Descartes, Euler, Pascal, Cantor, many more. Anecdotal, illuminating. 30 diagrams. Bibliography. 256pp. 5⅜ x 8½. 28973-7

HISTORY OF MATHEMATICS, David E. Smith. Nontechnical survey from ancient Greece and Orient to late 19th century; evolution of arithmetic, geometry, trigonometry, calculating devices, algebra, the calculus. 362 illustrations. 1,355pp. 5⅜ x 8½. Two-vol. set. Vol. I: 20429-4 Vol. II: 20430-8

A CONCISE HISTORY OF MATHEMATICS, Dirk J. Struik. The best brief history of mathematics. Stresses origins and covers every major figure from ancient Near East to 19th century. 41 illustrations. 195pp. 5⅜ x 8½. 60255-9

CATALOG OF DOVER BOOKS

Physics

OPTICAL RESONANCE AND TWO-LEVEL ATOMS, L. Allen and J. H. Eberly. Clear, comprehensive introduction to basic principles behind all quantum optical resonance phenomena. 53 illustrations. Preface. Index. 256pp. 5⅜ x 8½. 65533-4

QUANTUM THEORY, David Bohm. This advanced undergraduate-level text presents the quantum theory in terms of qualitative and imaginative concepts, followed by specific applications worked out in mathematical detail. Preface. Index. 655pp. 5⅜ x 8½. 65969-0

ATOMIC PHYSICS: 8th edition, Max Born. Nobel laureate's lucid treatment of kinetic theory of gases, elementary particles, nuclear atom, wave-corpuscles, atomic structure and spectral lines, much more. Over 40 appendices, bibliography. 495pp. 5⅜ x 8½. 65984-4

A SOPHISTICATE'S PRIMER OF RELATIVITY, P. W. Bridgman. Geared toward readers already acquainted with special relativity, this book transcends the view of theory as a working tool to answer natural questions: What is a frame of reference? What is a "law of nature"? What is the role of the "observer"? Extensive treatment, written in terms accessible to those without a scientific background. 1983 ed. xlviii+172pp. 5⅜ x 8½. 42549-5

AN INTRODUCTION TO HAMILTONIAN OPTICS, H. A. Buchdahl. Detailed account of the Hamiltonian treatment of aberration theory in geometrical optics. Many classes of optical systems defined in terms of the symmetries they possess. Problems with detailed solutions. 1970 edition. xv+360pp. 5⅜ x 8½. 67597-1

PRIMER OF QUANTUM MECHANICS, Marvin Chester. Introductory text examines the classical quantum bead on a track: its state and representations; operator eigenvalues; harmonic oscillator and bound bead in a symmetric force field; and bead in a spherical shell. Other topics include spin, matrices, and the structure of quantum mechanics; the simplest atom; indistinguishable particles; and stationary-state perturbation theory. 1992 ed. xiv+314pp. 6⅛ x 9¼. 42878-8

LECTURES ON QUANTUM MECHANICS, Paul A. M. Dirac. Four concise, brilliant lectures on mathematical methods in quantum mechanics from Nobel Prize–winning quantum pioneer build on idea of visualizing quantum theory through the use of classical mechanics. 96pp. 5⅜ x 8½. 41713-1

THIRTY YEARS THAT SHOOK PHYSICS: The Story of Quantum Theory, George Gamow. Lucid, accessible introduction to influential theory of energy and matter. Careful explanations of Dirac's anti-particles, Bohr's model of the atom, much more. 12 plates. Numerous drawings. 240pp. 5⅜ x 8½. 24895-X

ELECTRONIC STRUCTURE AND THE PROPERTIES OF SOLIDS: The Physics of the Chemical Bond, Walter A. Harrison. Innovative text offers basic understanding of the electronic structure of covalent and ionic solids, simple metals, transition metals and their compounds. Problems. 1980 edition. 582pp. 6⅛ x 9¼. 66021-4

CATALOG OF DOVER BOOKS

HYDRODYNAMIC AND HYDROMAGNETIC STABILITY, S. Chandrasekhar. Lucid examination of the Rayleigh-Benard problem; clear coverage of the theory of instabilities causing convection. 704pp. 5⅜ x 8¼. 64071-X

INVESTIGATIONS ON THE THEORY OF THE BROWNIAN MOVEMENT, Albert Einstein. Five papers (1905–8) investigating dynamics of Brownian motion and evolving elementary theory. Notes by R. Fürth. 122pp. 5⅜ x 8½. 60304-0

THE PHYSICS OF WAVES, William C. Elmore and Mark A. Heald. Unique overview of classical wave theory. Acoustics, optics, electromagnetic radiation, more. Ideal as classroom text or for self-study. Problems. 477pp. 5⅜ x 8½. 64926-1

PHYSICAL PRINCIPLES OF THE QUANTUM THEORY, Werner Heisenberg. Nobel Laureate discusses quantum theory, uncertainty, wave mechanics, work of Dirac, Schroedinger, Compton, Wilson, Einstein, etc. 184pp. 5⅜ x 8½. 60113-7

ATOMIC SPECTRA AND ATOMIC STRUCTURE, Gerhard Herzberg. One of best introductions; especially for specialist in other fields. Treatment is physical rather than mathematical. 80 illustrations. 257pp. 5⅜ x 8½. 60115-3

AN INTRODUCTION TO STATISTICAL THERMODYNAMICS, Terrell L. Hill. Excellent basic text offers wide-ranging coverage of quantum statistical mechanics, systems of interacting molecules, quantum statistics, more. 523pp. 5⅜ x 8½. 65242-4

THEORETICAL PHYSICS, Georg Joos, with Ira M. Freeman. Classic overview covers essential math, mechanics, electromagnetic theory, thermodynamics, quantum mechanics, nuclear physics, other topics. xxiii+885pp. 5⅜ x 8½. 65227-0

PROBLEMS AND SOLUTIONS IN QUANTUM CHEMISTRY AND PHYSICS, Charles S. Johnson, Jr. and Lee G. Pedersen. Unusually varied problems, detailed solutions in coverage of quantum mechanics, wave mechanics, angular momentum, molecular spectroscopy, more. 280 problems, 139 supplementary exercises. 430pp. 6½ x 9¼. 65236-X

THEORETICAL SOLID STATE PHYSICS, Vol. I: Perfect Lattices in Equilibrium; Vol. II: Non-Equilibrium and Disorder, William Jones and Norman H. March. Monumental reference work covers fundamental theory of equilibrium properties of perfect crystalline solids, non-equilibrium properties, defects and disordered systems. Total of 1,301pp. 5⅜ x 8½.   Vol. I: 65015-4   Vol. II: 65016-2

WHAT IS RELATIVITY? L. D. Landau and G. B. Rumer. Written by a Nobel Prize physicist and his distinguished colleague, this compelling book explains the special theory of relativity to readers with no scientific background, using such familiar objects as trains, rulers, and clocks. 1960 ed. vi+72pp. 23 b/w illustrations. 5⅜ x 8½.
42806-0 $6.95

A TREATISE ON ELECTRICITY AND MAGNETISM, James Clerk Maxwell. Important foundation work of modern physics. Brings to final form Maxwell's theory of electromagnetism and rigorously derives his general equations of field theory. 1,084pp. 5⅜ x 8½. Two-vol. set.   Vol. I: 60636-8   Vol. II: 60637-6

QUANTUM MECHANICS: Principles and Formalism, Roy McWeeny. Graduate student-oriented volume develops subject as fundamental discipline, opening with review of origins of Schrödinger's equations and vector spaces. Focusing on main principles of quantum mechanics and their immediate consequences, it concludes with final generalizations covering alternative "languages" or representations. 1972 ed. 15 figures. xi+155pp. 5⅜ x 8½. 42829-X

INTRODUCTION TO QUANTUM MECHANICS WITH APPLICATIONS TO CHEMISTRY, Linus Pauling & E. Bright Wilson, Jr. Classic undergraduate text by Nobel Prize winner applies quantum mechanics to chemical and physical problems. Numerous tables and figures enhance the text. Chapter bibliographies. Appendices. Index. 468pp. 5⅜ x 8½. 64871-0

METHODS OF THERMODYNAMICS, Howard Reiss. Outstanding text focuses on physical technique of thermodynamics, typical problem areas of understanding, and significance and use of thermodynamic potential. 1965 edition. 238pp. 5⅜ x 8½. 69445-3

TENSOR ANALYSIS FOR PHYSICISTS, J. A. Schouten. Concise exposition of the mathematical basis of tensor analysis, integrated with well-chosen physical examples of the theory. Exercises. Index. Bibliography. 289pp. 5⅜ x 8½. 65582-2

THE ELECTROMAGNETIC FIELD, Albert Shadowitz. Comprehensive undergraduate text covers basics of electric and magnetic fields, builds up to electromagnetic theory. Also related topics, including relativity. Over 900 problems. 768pp. 5⅝ x 8¼. 65660-8

GREAT EXPERIMENTS IN PHYSICS: Firsthand Accounts from Galileo to Einstein, Morris H. Shamos (ed.). 25 crucial discoveries: Newton's laws of motion, Chadwick's study of the neutron, Hertz on electromagnetic waves, more. Original accounts clearly annotated. 370pp. 5⅜ x 8½. 25346-5

RELATIVITY, THERMODYNAMICS AND COSMOLOGY, Richard C. Tolman. Landmark study extends thermodynamics to special, general relativity; also applications of relativistic mechanics, thermodynamics to cosmological models. 501pp. 5⅜ x 8½. 65383-8

STATISTICAL PHYSICS, Gregory H. Wannier. Classic text combines thermodynamics, statistical mechanics, and kinetic theory in one unified presentation of thermal physics. Problems with solutions. Bibliography. 532pp. 5⅜ x 8½. 65401-X

Paperbound unless otherwise indicated. Available at your book dealer, online at **www.doverpublications.com**, or by writing to Dept. GI, Dover Publications, Inc., 31 East 2nd Street, Mineola, NY 11501. For current price information or for free catalogs (please indicate field of interest), write to Dover Publications or log on to **www.doverpublications.com** and see every Dover book in print. Dover publishes more than 500 books each year on science, elementary and advanced mathematics, biology, music, art, literary history, social sciences, and other areas.